This book, written by one of philosophy's pre-eminent logicians, argues that many of the basic assumptions commonly made in logic, the foundations of mathematics and metaphysics are in need of change. It is therefore a book of critical importance to logical theory and the philosophy of mathematics.

Jaakko Hintikka proposes a new basic first-order logic and uses it to explore the foundations of mathematics. This new logic enables logicians to express on the first-order level such concepts as equi-cardinality, infinity and truth in the same language. The famous impossibility results by Gödel and Tarski that have dominated the field for the past sixty years turn out to be much less significant than has been thought. All of ordinary mathematics can in principle be done on this first-order level, thus dispensing with all problems concerning the existence of sets and other higher-order entities.

THE PRINCIPLES OF MATHEMATICS REVISITED

THE PRINCIPLES OF MATHEMATICS REVISITED

JAAKKO HINTIKKA

Boston University

CAMBRIDGE
UNIVERSITY PRESS

PUBLISHED BY THE PRESS SYNDICATE OF THE UNIVERSITY OF CAMBRIDGE
The Pitt Building, Trumpington Street, Cambridge, United Kingdom

CAMBRIDGE UNIVERSITY PRESS
The Edinburgh Building, Cambridge CB2 2RU, UK   http://www.cup.cam.ac.uk
40 West 20th Street, New York, NY 10011-4211, USA   http://www.cup.org
10 Stamford Road, Oakleigh, Melbourne 3166, Australia
Ruiz de Alarcón 13, 28014 Madrid, Spain

First published 1996
First paperback edition 1998
Reprinted 1999

Typeset in Times

Library of Congress Cataloging in Publication data is available.

A catalog record for this book is available from the British Library.

ISBN 0 521 49692 6 hardback
ISBN 0 521 62498 3 paperback

Transferred to digital printing 2003

Contents

Introduction   *page* vii

1 The functions of logic and the problem of truth
   definition   1
2 The game of logic   22
3 Frege's fallacy foiled: Independence-friendly logic   46
4 The joys of independence: Some uses of IF logic   72
5 The complexities of completeness   88
6 Who's afraid of Alfred Tarski? Truth definitions for
   IF first-order languages   105
7 The liar belied: Negation in IF logic   131
8 Axiomatic set theory: Fraenkelstein's monster?   163
9 IF logic as a framework for mathematical theorizing   183
10 Constructivism reconstructed   211
11 The epistemology of mathematical objects   235
Appendix (by Gabriel Sandu)   254

References   271
Index of names   281
Index of subjects and titles   285

Introduction

The title of this work is modeled on Bertrand Russell's 1903 book *The Principles of Mathematics*. What is the connection? As I see it, Russell's book was an important step in his struggle to liberate himself from traditional approaches to logic and the foundations of mathematics and to replace them by an approach using as its main tool, and deriving its inspiration from, the new logic created by Frege and Peano. In the *Principles*, Russell is not yet actually constructing the new foundation for mathematics which he later built with A. N. Whitehead. The *Principles* is not the *Principia*. What Russell is doing in the 1903 book is to examine the conceptual problems that arise in the foundations of logic and mathematics, expose the difficulties in the earlier views and by so doing try to find guidelines for the right approach.

In this book, I am hoping in the same spirit to prepare the ground for the next revolution (in Jefferson's sense rather than Lenin's) in the foundations of mathematics. As in Russell, this involves both a critical and a constructive task, even though they cannot be separated from each other. Indeed, if there had not been the danger of confusing bibliographers, I would have given this book the double entendre title *The Principles of Mathematics Revis(it)ed*.

The critical part of my agenda is the easier one to describe. Indeed, if I were Thomas Kuhn, I could give a most concise description of the state of the art in the foundations of logic and mathematics. Almost everyone in the field assumes that he or she is doing normal science, even though in reality a serious crisis is about to break out. Hence what is needed here is crisis science in Kuhn's sense, and not normal science with its safe, or at least generally accepted, conceptual framework and its agreed-upon criteria of success. Most of the received views concerning the overall situation in the foundations of logic and mathematics are not only open to doubt but are arguably (and often demonstrably) false. Indeed, arguments and demonstrations to this effect will be presented in this work.

The dogmas that are ripe for rejection include the following commonplaces: The basic part of logic, the true elementary logic, is ordinary first-order logic, at least if you are a classicist. If you are an intuitionist, you use instead Heyting's intuitionist first-order logic. In either case, our basic elementary logic admits of a complete axiomatization.

On the level of first-order logic, you can only formulate formal rules of inference, that is, treat logic syntactically. For (and with this point I agree) in order to do semantics (model theory), you need a truth definition for the language in question. According to the current views, such a truth definition cannot be formulated in the same language, only in a stronger metalanguage. For first-order languages, such a truth definition must naturally be formulated in set theory or in a second-order language. In both cases, we are dealing with a mathematical rather than a purely logical language. Model theory is thus inevitably a mathematical discipline, and not a purely logical one. Moreover, what a formal truth definition can do is only an abstract correlation between sentences and the facts that make them true. They cannot provide an explanation of what it is that makes a sentence true. Formal truth definitions cannot show, either, how sentences are actually verified.

Nontrivial first-order mathematical theories, including elementary arithmetic, are inevitably incomplete in all the interesting senses of the term. There are no absolute principles to guide the search for stronger deductive axioms for mathematical theories.

First-order logic is incapable of dealing with the most characteristic concepts and modes of inference in mathematics, such as mathematical induction, infinity, equicardinality, well-ordering, power set formation, and so forth. Hence mathematical thinking involves essentially higher-order entities of some sort or the other, be they sets, classes, relations, predicates, and so forth, in the strong sense of involving quantification over them. Formally speaking, mathematics can be done on the first-order level only if the values of individual variables include higher-order entities, such as sets. It is therefore theoretically illuminating to formulate mathematical theories in set-theoretical terms. Axiomatic set theory is accordingly a natural framework for mathematical theorizing.

Negation is a simple concept which involves merely a reversal of truth-values, *true* and *false*. It is a characteristic feature of constructivistic approaches to logic to deny the law of excluded middle. Another trademark of constructivism is the rejection of the axiom of choice. In general, the way to implement a constructivistic interpretation of logic is to change the rules of inference or perhaps to change some of the clauses in a truth definition.

Furthermore, it is generally (though not universally) assumed by linguists, logicians and philosophers that one should try to abide by the so-called principle of compositionality in one's semantics. It says that the semantical value of a given complex expression is always a function of the semantical values of its component expressions. The function in question is determined by the syntactical form of the given expression.

In this work, I will argue that not only one or two of these widespread views are mistaken, but that *all* of them are wrong. Moreover, I will show what the correct views are that should replace the erroneous dogmas listed above. I am not a deconstructivist or a skeptic. On the contrary, I am trying to wake my fellow philosophers of mathematics from their skeptical slumbers and to point out to them a wealth of new constructive possibilities in the foundations of mathematics. This will be done by developing a new and better basic logic to replace ordinary first-order logic. This received basic logic owes its existence to an unduly restrictive formulation of the formation rules for first-order logic probably motivated by an uncritical acceptance of the dogma of compositionality. The new logic which is free of these restrictions provides the insights that in my judgment are needed in the foundations of logic and mathematics. Ironically, the received first-order logic which I am demoting is the core part of the then new logic to whose introduction Russell was paving the way in the *Principles of Mathematics*. The admirers of Russell can nevertheless find some consolation in the closeness of my new logic to the kind of logic Russell was propagating. The two nevertheless differ significantly in their philosophical and other general theoretical implications and suggestions.

In my judgment, even the critical aspects of this book will have a liberating effect. The reason is that many of the dogmas I have just listed are essentially restrictive, including theorems as to what cannot be done in logic or in mathematics. Refuting them is therefore an act of liberation. For a significant example, it turns out that in one perfectly natural sense of the word mathematics can in principle be done on the first-order level. In general my liberated first-order logic can do much more in mathematics than philosophers have recently thought of as being possible.

The strategy of this logician's liberation movement does not rely on complicated technical results, but on a careful examination of several of the key concepts we employ in logic and mathematics, including the following concepts: quantifier, scope, logical priority and logical dependence, completeness, truth, negation, constructivity, and knowledge of objects as distinguished from knowledge of facts. A patient analysis of these ideas can uncover in them greater conceptual riches than philosophers and mathematicians have suspected. It has recently been fashionable to badmouth Carnap's idea of philosophy of science as a logical analysis of the language of science, and by analogy of the philosophy of mathematics as the logical analysis of the "language" of mathematics – that is, of the basic concepts of mathematics and metamathematics. Undoubtedly Carnap and his colleagues were neither sensitive enough to the deeper conceptual issues nor in possession of sufficiently powerful logical tools. Nonetheless, objectively speaking a study of the roles of language, truth and logic in mathematics seems to me by far the most fruitful approach to the philosophical nature of mathematical theories. For a brief while I was playing with the idea of calling this book *Language, Truth and Logic in Mathematics*.

This strategy implies that my book will be more like Russell's *Principles* than his *Principia* in another respect, too. Most of the time I am not developing my ideas here in the form of a mathematical or logical treatise, with full formal details and explicit proof. This work is a philosophical essay, and not a research paper or treatise in logic or mathematics. Even though I will try to explain all the main formal details that I need in this book, I will accordingly do so only in so far as they seem to be necessary for the purpose of understanding my overall line of thought.

In thinking about the subject matter – or, rather, subject matters – of this book over the years I have profited from exchanges with more people than I can recall, let alone list here. Three of them occupy a special position, however, because they have made a contribution of one kind or the other to the genesis of this book. Some of the points I am making here I have argued with – or, rather, against – one of my oldest American friends, Burton Dreben, ever since I came to know him in 1954. After forty years, I believe I have finally found the clinching arguments for my line of thought.

Most of the new ideas expounded in this book I have developed in close cooperation with Gabriel Sandu. In some cases, I cannot any longer say who came upon which idea first. I can only hope that he will soon publish his version of some of the main results of this book. Sandu has also contributed to this book substantially through specific comments, criticisms and suggestions. He has written an important comparison between some of our results and Kripke's approach to truth which is printed below as an appendix to this book. In brief, this book owes more to Gabriel Sandu than I am likely to be aware of myself. Furthermore, Janne Hiipakka has been of invaluable help, not only in preparing the manuscript but also in correcting mistakes and making suggestions.

In working on this book, I have come to appreciate more and more the importance of the issues raised by the classics of twentieth-century philosophy of mathematics, especially Hilbert, Tarski, Carnap and Gödel. While I frequently disagree with each of them, the questions they were asking are much more important than most of the issues that have been debated recently, and eminently worth going back to for problems and ideas. I only regret that I have developed my own ideas far too late to discuss them with the likes of Tarski, Carnap or Gödel.

I have not received any direct grant support or release time while writing this book. Indirect support includes a grant from the Academy of Finland which has made possible the work of Sandu and Hiipakka. Travel grants from the Academy of Finland have also facilitated several of my working trips to Finland. Boston University has provided a congenial working environment. In fact, the origins of this book can be traced back to the insight that self-applied truth definitions are possible for independence-friendly languages; an insight which first occurred to me in the midst of a seminar I was conducting at Boston University in 1991. I have also had an opportunity to help organize several symposia under the auspices of the Boston Colloquium for Philosophy and History of Science which have provided stimuli for the ideas that have reached fruition in this book.

Relatively little of the material published here has been available earlier. Much of Chapter 10 has been published under the title "Constructivism *Aufgehoben*" in the proceedings of the "Logica 94" conference held in the Czech Republic in June 1994. A version of Chapter 2 has appeared in *Dialectica* vol. 49 (1995), pp. 229–249 under the title, "The Games of Logic and the Games of Inquiry." Some examples and some of the other material in this book, especially in Chapter 7, are lifted from the preprint entitled *Defining Truth, the Whole Truth and Nothing But the Truth*, published in the preprint series of the Department of Philosophy, University of Helsinki, in 1991. The rest of my literary "thefts" from my own earlier works are unintentional.

Finally my thanks are due to Cambridge University Press for accepting my book for publication in their program. It is a special compliment for a work on the foundations of mathematics to share a publisher with Russell and Whitehead.

1

The Functions of Logic and the Problem of Truth Definition

The most pervasive misconception about the role of logic in mathematical theorizing may turn out to be the most important one. Admittedly, this mistake is easily dressed up to look like a biased emphasis or an exaggerated division of labor. But it is nonetheless a serious matter. It can be exposed by asking the naive-sounding question: What can logic do for a mathematician? What *is* the role of logic in mathematics?

As a case study, I propose to consider one of the most influential works in the foundations of mathematics and one of the last major works that does not use the resources of modern logic. This work is Hilbert's *Foundations of Geometry* (1899). What Hilbert does there is to present an axiomatization (axiom system) of Euclidean geometry. Such a system is a *nonlogical* axiom system. It is a systematization of the truths (ordinary scientific or mathematical truths, not logical truths) of some discipline, usually some branch of mathematics or science. The systematization is obtained by, as it were, compressing all the truths about the subject matter into a finite (or recursively enumerable) set of axioms. In a certain sense, they are supposed to tell you everything there is to be told about this subject matter. Such an axiomatization, if complete, will give you an overview of the entire field in question. If you have reached in your investigation into this field a complete axiom system, then the rest of your work will consist in merely teasing out the logical consequences of the axioms. You do not any longer need any new observations, experiments or other inputs from reality. It suffices to study the axioms; you no longer need to study the reality they represent. Obviously, this intellectual mastery of an entire discipline is the most important attraction of the entire axiomatic method.

Philosophers sometimes think of the axiomatic method as a way of justifying the truths that an axiom system captures as its theorems. If so, the axioms have to be more obvious than the theorems, and the derivation of theorems from axioms has to preserve truth. The latter requirement is discussed below. As to the former, the obviousness requirement plays no role in the most important scientific theories. No one has ever claimed that Maxwell's or Schrödinger's equations are intuitively obvious. The interest of such fundamental physical equations is not even diminished essentially by the knowledge that they are only approximately true. The explanation is that such equations still offer an overview of a wide range of phenomena. They are means of the intellectual mastery of the part of reality they deal with. Generally speaking, this task of intellectual mastery is a much more important motivation of the axiomatic method than a quest of certainty.

Hilbert's axiomatization of geometry is an extreme example of this fact. Hilbert does not even raise the question whether the axioms of Euclidean geometry are true in actual physical space. All he is interested in is the structure delineated by the axioms. This structure is spelled out by the theorems of the axiom system. Whether this structure is instantiated by what we call points, lines and planes or by entities of some other kind is immaterial to his purpose.

But what is the job of logic in this enterprise? In Hilbert's treatise it remains tacit or at least unformalized. Not a single symbol of formal logic disfigures Hilbert's pages. Hilbert's monograph could have been written even if Boole, Frege and Cantor had never been born. Yet logic is an all-important presence in Hilbert's treatise in more than one way.

One way is completely obvious to any contemporary philosopher, logician or mathematician. In fact, this role of logic is so obvious that subsequent thinkers tend to take it for granted, thereby missing much of the historical significance of Hilbert's achievement. Hilbert is envisaging what is nowadays called a purely logical axiom system. That is to say, all substantive assumptions are codified in the axioms, whereas all the theorems are derived from the axioms by purely logical means. This idea may be taken to be the gist of the entire axiomatic method. Familiarity with this idea is so complete in our days that we easily forget what a bold novelty such a treatment was in Hilbert's historical situation. One index of this novelty is the label that Hilbert's contemporaries put on his approach, calling it "formalistic". Of course this term is completely undeserved. When Hilbert said that instead of points, lines, and circles he could have spoken of chairs, tables, and beermugs, he was merely highlighting the purely logical nature of the derivation of theorems from axioms.[1] Because of the purely logical nature of Hilbert's axiom system, all proofs from axioms could be represented in an explicit formal (but of course interpreted) logical notation, without any relevant difference. This idea of a purely logical consequence was familiar to Aristotle, though not to such intervening thinkers as Kant. This purely logical character of the passage from the axioms to theorems does not mean that the axioms and the theorems must be uninterpreted. It does not mean that the axioms cannot be true (materially true). In fact, Hilbert elsewhere (1918, p. 149) indicated quite explicitly that in any actual application of geometry, the truth of the axioms is a material (empirical) matter. Even the actual truth of the axioms of continuity has to be empirically ascertained.[2]

A further distinction is nevertheless in order. It is important to realize that a nonlogical axiom system can be either interpreted, as in applied geometry or thermodynamics, or uninterpreted, as in set theory or lattice theory. The only difference between the two is that in the latter the fundamental nonlogical concepts are left uninterpreted. This does not make any difference to the derivation of consequences (e.g., theorems) from premises (e.g., axioms). Such derivations proceed exactly in the same way in the two cases, as long as they are purely logical. Hence in purely logically formulated nonlogical axiom systems the contrast between interpreted and uninterpreted systems does not matter. Hilbert's provocative slogan about the geometry of chairs, tables, and beermugs can be understood as flaunting this parity of interpreted and uninterpreted axiom systems when it comes to the derivation of theorems, and hence indirectly illustrating the purely logical character of geometrical proofs.

The same point can be put in a slightly different way. The question whether logical inferences can be captured by completely formal (computable, recursive) rules is independent of the question whether the language in which those inferences are drawn is "formal", that is, uninterpreted, or "informal", that is, interpreted. The question of interpretation pertains to the nonlogical constants of the language. Logical constants are assumed to have the same meaning in either case.

Uninterpreted nonlogical axiom systems may be thought of as pertaining to certain structures as such, while the corresponding interpreted systems deal with the actual instantiations of these structures.

Here, then, is an example of the role of logic which understandably is on top of the minds of most of my fellow philosophers, logicians and mathematicians. Logic is the study of the relations of logical consequence, that is, of relations of implication or entailment. Its concrete manifestation is an ability to perform logical inferences, that is, to draw deductive conclusions. I will call this the *deductive* function of logic.

The tools needed for this job are normally thought of as being collected into sundry *axiomatizations of logic*. The term "axiomatization" as used here is nevertheless a courtesy title. For a so-called axiomatization of some part of logic is merely a method of enumerating recursively all the logical truths expressible in some explicit ("formal") language or the other. This enumeration is usually made to appear uniform with nonlogical axiom systems, like Hilbert's system of geometry. That is to say, the enumeration is obtained by putting forward completely formal axioms to start the enumeration from, and equally formal rules of inference whose repeated applications are the means of accomplishing the enumeration. In spite of this similarity in expositional style, one must be aware of the fundamental differences between nonlogical and logical axiom systems. Nonlogical axiomatization trades in ordinary nonlogical (material or mathematical) truth, logical axiomatization in so-called logical truth. Nonlogical axiom systems normally have an intended interpretation, even when the derivations from axioms are purely logical, whereas an axiomatization of logic has to be purely formal to generate the mechanical enumerability. A nonlogical axiom system can be formulated without saying anything about any explicitly formulated logic. In fact, this is what Hilbert does in the *Grundlagen* (1899). Even though all his derivations of theorems from axioms are carried through in conformity with impeccable logic, not a single explicit rule of logical inference is ever appealed to by Hilbert in his famous book.

From an axiomatization of logic, patterns of valid logical *inference* are obtained as a special case in the form of conditionals

$$S_1 \supset S_2 \tag{1.1}$$

If and only if such a *conditional* (with S_1 as the antecedent and S_2 as the consequent) is logically true (valid, as a sometime alternative term goes) is an *inference* from S_1 to S_2 valid.

Nevertheless in the course of this book I will point out difficulties with this characterization of valid inferences in terms of logical truth. I will return to this matter at the end of Chapter 7.

It is important to realize that this task of spelling out the valid inference-patterns cannot be entrusted to the ill-named "rules of inference" of an axiomatization of logic. For the raison d'être of such "rules of inference" is to serve as a vehicle for enumerating logical truths, and not material or mathematical truths. Hence the only thing required of them is that they preserve *logical* truth. For this purpose, it is not even necessary that they preserve material truth, that is, truth *simpliciter*. There are in fact plenty of examples of so-called inference rules used in the axiomatization of some part of logic that do not preserve truth as such. The best known cases in point are probably the rules of necessitation in modal logic. In them, one may for instance "infer" the logical truth of

$$N(S_1 \supset S_2) \tag{1.2}$$

from the logical truth of

$$S_1 \supset S_2 \tag{1.3}$$

where N is the necessity operator. But of course from the plain (natural) truth of (1.3) one cannot infer the truth of (1.2). Similar examples can also be found in ordinary nonmodal logic.

Thus a philosopher should be extra careful with such terms as "rule of inference" or even "rule of logical inference". They can in the wrong hands be viciously ambiguous. At the very least, we have to distinguish carefully between a logical proof (from premises) of material truth and logical (pseudo) proof of a logical truth. The latter is best looked upon merely as a segment of the recursive enumeration of logical truths.

For a while, mathematicians tended to take the basic rules of valid logical inference for granted. Sometimes they were thought of as being too obvious to need any particularly explicit treatment. Sometimes it was assumed – and is still being assumed – that Frege and Russell accomplished this task once and for all. Or if they did not, then Hilbert's henchman Wilhelm Ackermann did so. For what

Ackermann did in 1928 under Hilbert's supervision was to formulate for the first time that basic ground-floor part of logic (as it seems to be) that everybody and his sister have taken for granted ever since. This is what is usually known as first-order logic, quantification theory or (lower) predicate calculus. It used to be presented as if it resulted from a minor regimentation of our ordinary-language use of *everys* and *somes*. This idea seems to be vindicated by Chomsky's use of quantificational logic as the main medium of his born-again logical forms (LF's) of natural-language sentences (see, e.g., Chomsky 1977, p. 59). Small wonder, therefore, that first-order logic is generally taken to be the safe, unproblematic core area of logic, and that the languages relying on it are considered as a natural vehicle of our normal reasoning and thinking. A number of concrete developments in the foundations of mathematics in fact prompted logicians and mathematicians to present explicit axiomatizations of this part of logic separate from those higher echelons of logic which involve quantification over such abstract entities as predicates, classes, sets, and/or relations.

It is in fact easy to see why first-order logic at first seems to be a logician's dream come true. First and foremost this logic admits of a complete axiomatization. This seems at first sight to effectively fulfill the hopes of Hilbert and others that there is an unproblematic and completely axiomatizable basic logic. The existence of such a logic was in fact one of the presuppositions of what is known as Hilbert's program – a program, that is, of proving the consistency of certain mathematical theories, primarily arithmetic and analysis, by showing that one cannot formally derive a contradiction from their axioms. If the logic that is being used is not complete, then the entire project loses its point for there may then be unprovable contradictions among the logical consequences of the axioms. Luckily, as it seems, first-order logic was shown to be completely axiomatizable by Gödel in 1930. Furthermore, first-order logic can be shown to admit all sorts of nice metalogical results, such as compactness (an infinite set of sentences is consistent if and only if all its finite subsets are), upwards Löwenheim–Skolem theorem (a consistent finite set of sentences has a countable model), separation theorem (if σ and τ are consistent sets of formulas but $\sigma \cup \tau$ is inconsistent, then for some "separation formula" S in the shared vocabulary of σ and τ we have $\sigma \vdash S$, $\tau \vdash \sim S$), interpolation theorem (if $\vdash (S_1 \supset S_2)$ nontrivially, then for some formula I in the shared vocabulary of S_1 and S_2,

$\vdash (S_1 \supset I)$, $\vdash (I \supset S_2))$, Beth's theorem (implicit definability implies explicit definability), and so forth. In brief, first-order logic does not only seem basic, it is very nearly looks like a logician's paradise.

Admittedly, constructivistic logicians and mathematicians have sought to change some of the rules of first-order logic. But this does not seem to have shaken most philosophers' beliefs that in some size, shape or form first-order logic is the true unproblematic core area of logic.

Philosophically, first-order logic owes its special status largely to the fact that it is in a sense a nominalistic enterprise. It involves quantification only over individuals, that is, particular objects or entities that are dealt with as if they were particulars. This is in fact why this logic is called first-order logic.

However, storm clouds begin to gather as soon as logicians venture beyond the enchanted land of first-order logic. And they have to do so, for unfortunately first-order logic soon turns out to be too weak for most mathematical purposes. Its resources do not suffice to characterize fully such crucial concepts as mathematical induction, well-ordering, finiteness, cardinality, power set, and so forth. First-order logic is thus insufficient for most purposes of actual mathematical theorizing. Moreover, the terrain beyond it has proved uncharted and treacherous. It is notorious that those logical methods which go beyond first-order logic give rise to serious problems, no matter whether we are dealing with set theory or higher-order logic (type theory). There does not at first seem to be much to choose between the two. Second-order logic has been branded by Quine as set theory in sheep's clothing. And there is in fact a multitude of overlapping problems concerning the two. In highlighting these problems, philosophers of yore used to point to paradoxes sometimes amounting to outright contradictions. The sad truth is that even if we rid ourselves of all threatening inconsistencies, maybe by some sort of type stratification (as in second-order logic and more generally in higher-order logic), we still have the problem of finding out which formulas are logically true. Eliminating what is logically false does not give us a means of ascertaining what is logically true. Often these problems are quite specific. Is the axiom of choice valid? What axioms do we need in axiomatic set theory? How are higher-order quantifiers to be understood? The list of serious problems can be continued ad nauseam, if not ad infinitum.

If I ended my survey of the role of logic in mathematical theories here, perhaps with some further elaboration of the difficulties connected with the uses of logic beyond first-order theories, few philosophers would be surprised. However, such a procedure would leave half of the real story untold, and indeed the more basic half of the story of the uses of logic in mathematics.

In order to enable you to see what I mean, I can ask: What would you have to do in order to turn Hilbert's axiomatization of geometry into a state of the art logic-based axiom system? What I have discussed so far is what has to be done to Hilbert's argumentation, that is, to the lines of reasoning which lead him from the axioms to the theorems. Most of such reasoning turns out to utilize first-order reasoning, with a smattering of number-theoretical and set-theoretical reasoning thrown in. But even before we can hope to express Hilbert's inferences in the language of logic, we have to express in such a language the ultimate premises of his inferences that is to say, the axioms of geometry. And here is the other, unjustly underemphasized, second function of logical concepts in mathematics. They are relied on essentially in the very formulation of mathematical theories. We can make the axioms of a typical mathematical theory say what they say only by using suitable logical concepts, such as quantifiers and logical connectives.

This fact is vividly illustrated once again by Hilbert's treatment of geometry. His axioms deal with certain specified relations – between-ness, equidistance, and so forth, – between certain specified kinds of objects – points, lines, and so forth. What the axioms say about these entities is formulated by means of the basic logical notions, mainly quantifiers and propositional connectives. Of course, Hilbert does not use any explicit logical symbols, but formulates his axioms by means of ordinary mathematical language. But anyone who has a modicum of elementary logic can write down a first-order formulation of all of Hilbert's axioms in fifteen minutes, with two exceptions. They are the Archimedean axiom, which relies on the notion of natural number, and the so-called axiom of completeness, which turns on the notion of the maximality of a model.

The Archimedean axiom says that from a given line segment one can reach any given point on the same line by extending the segment by its own length a finite number of times. The axiom of complete-ness was a second thought on Hilbert's part. It appears first in Hilbert (1900), and makes its entry into Hilbert's geometrical monograph in its French translation (1900) and then in the second German edition (1903). Its import can be expressed by saying that the intended models of the other axioms must be maximal in the sense that no new geometrical objects can be added to them without making one of these axioms false.

More generally, much of the foundational work that has been done since Cauchy by working mathematicians consisted in expressing in first-order logical terms the precise contents of different mathematical concepts. A familiar but instructive case in point is offered by the so-called ε-δ definitions of concepts like continuity and differentiation. (I will return to them in Chapter 2.) Another example is found in the different and increasingly general concepts of integral that were developed in the late nineteenth and early twentieth centuries. In developing such ideas, mathematicians were not engaged in the discovery of new mathematical truths. They were engaged in analyzing different mathematical concepts in logical terms.

I will call this function of logic (logical concepts) in expressing the content of mathematical propositions its *descriptive* function. Many interesting phenomena in the foundations of mathematics become easier to understand in the light of the tension which there often is in evidence between this descriptive function and the deductive function of logic in mathematics. Arguably, the descriptive function is the more basic of the two functions I have distinguished from each other. If mathematical propositions were not expressed in terms of logical concepts, their inferential relationships would not be possible to handle by means of logic.

What I have called the descriptive function of logic can be put into service as a tool of conceptual analysis. This possibility is illustrated by examples from the history of mathematics like the ones just mentioned, but it is not restricted to mathematics. It is what underlies Hilbert's less than completely happy terminology when he calls his axioms implicit definitions of the geometrical concepts involved in them.

It is important to realize that this descriptive function of logic in formulating mathematical axioms is the same – and hence equally indispensable – no matter whether the axioms belong to an interpreted axiom system, for instance, to the axioms of thermodynamics or interpreted geometry, or to an uninterpreted axiom system, such as the axioms of group theory, theory of fields, or lattice theory. Even abstract mathematical theories can be thought of as explications of certain intuitive concepts, topology as an explication of the concept of continuity, group theory of the idea of symmetry, lattice theory of the notions related to the idea of ordering, and so forth. In all these cases, for the explicatory purpose logical notions of some sort are clearly a must. Thus we can see that those philosophers who maintain that logical notions like quantifiers do not have the same sense in mathematical theories as they have in everyday life are not only off the mark but off the map.[3] On the contrary, the representation of mathematical propositions in formal or informal language is predicated on the assumption that logical constants are being used with their normal meaning.

A realization of the difference between the two functions of logic in mathematics prompts further questions. The main one concerns the way different parts and aspects of logic serve the two (or three) functions. Which parts serve which function? What is required of logic that it can serve one or the other of the two functions? Or, must there for some reason be just one indivisible logic serving both the descriptive and the deductive purpose? These are fundamental questions in the philosophy of mathematics, and yet they have been scarcely raised in earlier literature. I will return to them at the end of Chapters 9 and 10.

Acknowledging the descriptive function of logic in mathematical theories in fact occasions an addition to what was said earlier of the purely logical character of an axiom system like Hilbert's *Grundlagen*. Not only must it be the case that all the proofs of theorems from axioms are conducted purely logically; it must also be required that the representational task of the axioms is accomplished by purely logical means. In practice this normally means that the only nonlogical notions in the axiom system are certain undefined properties and relations among the objects of the theory in question. Of them, only such things are assumed that are explicitly stated in the axioms.

This purely logical character of the formulation of axioms is only partly necessitated by the requirement that proofs of theorems from axioms must be purely logical. It is part and parcel of Hilbert's conception of the axiomatic method. A partial reason is that the only entities that the axioms may mention must be the objects of the theory. For instance, in the axioms of geometry we must not postulate a correspondence between points on a line and real numbers, for

THE FUNCTIONS OF LOGIC

the latter are not objects of geometry. In actual historical fact, Hilbert's choice of his axioms seems to have been deeply influenced by the requirement that they must be purely logical in the sense indicated here.

The neglect of the descriptive function of logical is especially striking in recent discussions of the philosophical problems of cognitive science. A central role is played there by the notion of representation, especially mental but to some extent also linguistic. Now, what would a typical situation of linguistic representation look like? On the basis of what was said earlier, representation by means of first-order logic should certainly be the test case. There may not be any completely presuppositionless Adamic discourse to serve as a testing-ground of theories of representation, but in the first-order heirs to Frege's *Begriffsschrift* we seem to have – *Gott sei Lob* – a simple mode of representation which everybody admits is a basic one. Or is it the case that we have to exempt a great many cognitive scientists and their pet philosophers from the scope of that "everybody"? In reading the relevant literature, it is hard to avoid the impression that many contemporary philosophers of cognitive science are desperately trying to restrict their attention to such forms of representation as do not involve logic. If so, we are witnessing a comedy in which one fraction of philosophers scrupulously tries to avoid discussing the very cases of representation which others treat as the paradigmatic ones. Or, is it perhaps that philosophers of cognition have not acknowledged the descriptive function of logic, which amounts in effect to its representative function? If so, they are disregarding the medium of representation for the overwhelming majority of our advanced theoretical knowledge.

And even if a philosopher believes that the all-important mental information processing uses some means other than linguistic or logical representation, he or she should still offer a realistic alternative account of the cognitive processes that can be carried out by means of first-order languages.

The systematic study of the deductive function of logic is known as proof theory. The systematic study of the descriptive function of logic is known as model theory or logical semantics. Several highly influential logicians and philosophers used to maintain – and in some cases still maintain–the impossibility of model theory as a large-scale philosophically relevant systematic enterprise. Some used to deny the very possibility of model theory outright, some its possibility in the all-important case of our actual working (or perhaps better, thinking) language, while still others merely deny its philosophical relevance. In different variants, such doubts have been expressed by Frege, Russell, Wittgenstein, the Carnap of the early thirties, Quine and Church. The way in which this tradition was gradually and partially overcome is studied briefly in Hintikka (1988b). This underestimation of model theory among philosophically oriented logicians can typically be traced back to a failure to appreciate the descriptive (representational) function of logic.

Notwithstanding such doubts, model theory has apparently proved possible, because actual, as the scholastics used to say. But questions can still be raised about its philosophical relevance. What precisely are the conceptual presuppositions of model theory? The crucial concept in any such theory is, unsurprisingly, the concept of model. The idea is to discuss what a sentence S says by associating with it a class of structures, also known as models, scenarios, systems, possible worlds, worlds, or what not. Let us call this class $M(S)$. Strictly speaking, there are two related senses of model here, depending on whether in the passage from S to $M(S)$ the nonlogical constants of S are allowed to be reinterpreted. Models of the second kind are of course the same as the models of the first kind *modulo* isomorphism. In the foundations of mathematics, the distinction between the two kinds of models makes little difference.

But how is $M(S)$ specified? In order to specify it, we must obviously do two things. First, we have to be given some class (set, space) Ω of models, that is, structures of the appropriate sort. Second, by reference to S we have to give the criteria as to when a given member M of Ω qualifies as one of the models of S.

The former question (the question of the choice of Ω) did not attract much attention on the part of logicians until the development of abstract (alias model-theoretical) logics (see here Barwise and Feferman 1985). Independently of them, certain interesting candidates for a nonstandard choice of Ω have meanwhile been proposed. They can involve on the one hand novel structures that were not earlier considered as models, such as the urn models of Veikko Rantala (1975) that will be explained in Chapter 5. On the other hand they can involve deliberate restrictions on the space of models. One especially interesting restriction is obtained by requiring that the surviving models should possess certain suitable extremality

THE FUNCTIONS OF LOGIC

THE FUNCTIONS OF LOGIC

THE FUNCTIONS OF LOGIC

(maximality and minimality) properties (see Hintikka 1993b). More work is nevertheless in order in this direction. Perhaps the "special models" just mentioned could in the future serve usefully as test cases in the very, very abstract theory of abstract logics.

I will return briefly to the ideas of maximality and minimality in Chapter 9.

For the purposes of my line of thought here, the central question is the second one, the question as to when a model M is a model of a sentence S. Here a partial answer is obvious. In the basic sense of model, M is a model of S if and only if S is *true* in M. And the conditions of a sentence being true in a model is what *truth definitions* codify. Hence the specification of the all-important relations of *being a model of* is essentially a matter of *truth definitions*.

The question of the possibility of truth definitions, and the question of the presuppositions of such truth definitions, are thus of major interest to the entire foundation of logic and mathematics. The philosophical viability of model theory stands or falls with the philosophical viability of truth definitions, and the relative dependence or independence of model theory on other approaches to foundations is a matter of whether those other approaches are presupposed in the relevant truth definitions. Even though Tarski himself never highlighted the fact (as far as I know), it is no accident that the same thinker developed both the first explicit truth definitions and later (with his students and associates) the contemporary model theory in the narrower technical sense.

The most commonly used type of truth definition was introduced by Alfred Tarski in 1935 – or perhaps 1933, if we heed the publication date of the Polish original. As will be explained more fully in Chapter 5, the guiding principle of Tarski's enterprise was what logicians at the time (and later) usually called the idea of recursive definition. Later linguists are wont to call it compositionality. This principle says that the semantic attributes of a complex expression are functions of the semantic attributes of its constituent expressions. Applied to truth definitions, it says that the truth-value of a sentence is determined by the semantical attributes of its constituent expressions. One perhaps hopes to be able to say here: it depends only, on the truth-values of its constituent expressions. Unfortunately the constituent expressions of quantified sentences typically contain free variables. They are open formulas, not sentences, and therefore cannot have a truth-value.

This explains the first main feature of Tarski's truth definition. The truth-value of a sentence is defined by him with the help of another notion which applies also to open formulas, namely, the notion of satisfaction. The definition is applied from inside out, beginning with the simplest (atomic) formulas in the sentence in question. A crucial role is played by the notion of a valuation, which is indeed an assignment of individuals to each individual constant and each individual variable of the language in question as its values. This reliance of Tarski-type definitions of truth on the auxiliary concept of satisfaction has, strangely enough, led some scholars to deny that Tarski's truth definition is compositional.

Very briefly, a Tarski-type truth definition can be explained by keeping in mind that truth is relative to a model M and a valuation v. A valuation assigns to each nonlogical primitive symbol, including the individual variables $x_1, x_2, \ldots, x_i, \ldots$, an entity of the appropriate type from the model M. A sentence (closed formula) is true if and only if there is a valuation that satisfies it. Satisfaction is defined recursively in the obvious way. For instance, $(\exists x_i)S[x_i]$ is satisfied by a valuation v if and only if there is a valuation which differs from v only for the argument x_i and which satisfies $S[x_i]$. Likewise, v satisfies $(\forall x_i)S[x_i]$ if and only if every valuation that differs from v only on x_i satisfies $S[x_i]$. For propositional connectives, the usual truth-table conditions are used to characterize satisfaction.

The satisfaction of an atomic formula is defined in the usual way. For instance, $R(x_i, x_j)$ is satisfied by v if and only if $\langle v(x_i), v(x_j) \rangle \in v(R)$. Putting all these stipulations together, we can indeed arrive at a recursive truth definition.

The further details of Tarski's procedure do not have to concern us here, but only the main features of his truth definition. If a truth definition is formulated explicitly in a metalanguage, it is natural to assume that that metalanguage contains elementary arithmetic. Then one can use the normal technique of Gödel numbering to discuss the syntax of the first-order language in question. If this language contains a finite number of predicate and function symbols, then the logical type of the valuation function v is essentially a mapping from natural numbers (Gödel numbers of symbols and formulas) into the individuals of the domain do(M) of the model in question. The truth predicate itself which emerges from a Tarski-type treatment therefore has what logicians call a Σ_1^1 form. In other words, it has the form of a second-order existential quantifier (or a finite string of such quantifiers) followed by a first-order formula. All the quantifiers in this formula range over either natural numbers or else over the (other) individuals in the domain do(**M**) of the model in question. In other words, they are first-order quantifiers.

Indeed, we can see the most important feature of Tarski-type truth definitions for first-order languages: the definition itself is formulated in a second-order language, that is to say, in a language in which one can quantify over valuations. This feature of Tarski's truth definition was not an accidental one, as Tarski himself proved. Indeed, he proved that, given certain assumptions, a truth definition can be given for a language only in a stronger metalanguage. This is Tarski's famous impossibility result. It is closely related to Gödel's (1931) incompleteness results. Indeed, it has been established that Gödel first arrived at his incompleteness results by discovering the undefinability of arithmetical truth in a first-order arithmetical language.

The details of Tarski-type truth definitions and of Tarski's impossibility result are familiar from the literature, and therefore will not be elaborated here (cf. Ebbinghaus, Flum, and Thomas 1984, Ch. 3; Mostowski 1965, Ch. 3).

Truth definitions are important for several different reasons, though those reasons are not all equally directly relevant to the foundations of mathematics. Without the notion of truth there is little hope of capturing such basic concepts of logic as validity (truth in every model) and logical consequence. Furthermore, a truth definition is calculated to specify the truth-conditions of different sentences S. These truth-conditions are closely related to the notion of meaning, in the sense of sentence meaning. For what our assertively uttered sentence S in effect says can be paraphrased by saying that what S asserts is that its truth-conditions are satisfied. Thus to know the truth-conditions of S is to know what S means.

The only major qualification needed here has already been made. Truth-conditions deal with sentence meaning, not symbol meaning. The latter has to be treated separately, and it must be taken for granted in the formulation of truth-conditions and truth definitions. This point is useful to keep in mind. Some philosophers have, for instance, tried to criticize Tarski-type truth definitions for illicit reliance on the concept of meaning. The reliance is there, but it is not illicit because it is not circular. Tarski's project is nothing more and nothing less than to define truth-conditions in terms of symbol meaning, that is, sentence meaning in terms of symbol meaning. There is nothing wrong in such an attempt.

Tarski's and Gödel's negative results have been taken to have major philosophical and other general theoretical implications. The undefinability of such basic metalogical concepts as truth, validity and logical consequence on the first-order level shows that ordinary first-order logic is in an important sense not self-sufficient. Different philosophers have sought to draw different conclusions from this failure. Some rule out metalogic as a purely logical discipline, maintaining that it belongs to set theory and hence to mathematics. Others, like Putnam (1971), argue that logic must be taken to comprehend also higher-order logic and perhaps, by implication, parts of set theory. Generally speaking, Tarski's result seems to confirm one's worst fears about the dependence of model theory on higher-order logic and thereby on questions of sets and set existence. It has even been alleged that this makes model theory little more than a part of set theory. Indeed, the apparent dependence of Tarski-type truth definitions on set theory is in my view one of the most disconcerting features of the current scene in logic and in the foundations of mathematics. I am sorely tempted to call it "Tarski's curse". It inflicts model theory with all the problems and uncertainties of the foundations of set theory. More generally, Tarski's curse might be understood as the undefinability of truth for a given language in that language (given Tarski's assumptions). The importance of this negative result cannot be exaggerated. One of its first victims was Carnap's grandiose vision of a single universal language in which Hilbert's and Gödel's formalization techniques would also enable us to discuss its own semantics. In a more general perspective, Tarski's undefinability result inevitably gives every model theorist and most semanticists a bad intellectual conscience. The explicit formal languages Tarski's result pertains to were not constructed to be merely logicians' playthings. They were to be better tools, better object languages for the scientific and mathematical enterprise. But if a model theorist decides to study one of them, he or she will then have either to use a stronger metalanguage for the purpose or else to leave the metalanguage informal. In the former case, we have the blind leading the blind or, more specifically, the semantics of a language being studied by means of a more mysterious language, while in the latter case the semanticist has simply given up his or her professional responsibilities.

Second, one can try to apply Tarski's result to our actual working language, called "colloquial language" by Tarski himself. The application is not unproblematic, for Tarski's theorem is formulated so as to deal only with formal (but interpreted) languages satisfying certain explicit conditions. But assuming that the conditions of Tarski's result are satisfied by our ordinary working language, then we cannot define truth for this language. The main characteristic of our own actual language, duly emphasized by Tarski, is its universality. There is therrefore no stronger metalanguage beyond (or over) it in which the notion of truth for this universal language could be defined.

Hence presumably the assumptions on which Tarski's theorem rests do not apply to natural languages. But if so, we have reached a major negative metatheorem concerning what can be done by way of an explicit semantical theory of ordinary language. And even if we consider, instead of our everyday ("colloquial") discourse, the conceptual system codified in the language of science, we still obtain remarkable negative results. As Tarski himself expressed it on one occasion (Tarski 1992, in a letter to Neurath, 7 September 1936):

> There still remains the problem of universal language. It appears to me, that this problem has been completely cleared up through the discussions of the Varsovians (Lesniewski and myself) and also the Viennese (of Gödel and Carnap): one cannot manage with a universal language. Otherwise one would have to forego [sic] introducing and making precise the most important syntactical and semantical concepts ("true", "analytic", "synthetic", "consequence", etc.). One might now think that this circumstance is of no special importance for the actual sciences, that therefore for the purposes of actual sciences a single universal language can suffice entirely. Even this opinion appears to me as incorrect, and for the following reason in particular: to pursue actual sciences, something like physics, one must have available an extended mathematical apparatus; now we know however, that for every language (therefore also for the presumed "universal language") one can give entirely elementary number-theoretic concepts, respectively sentences, which in this language cannot be made precise, respectively cannot be proved.

More generally, the undefinability of truth can be considered a paradigmatic instance of the important but frequently unacknowledged assumption of the *ineffability of semantics*. This is not an occasion to examine the full role of this assumption in the general philosophy of the last hundred-odd years. Some aspects of its career are examined in Hintikka (1988b) and in Hintikka and Hintikka (1986, Ch. 1). However, it is relevant to note that this assumption dominated the early development of contemporary logical theory. It was embraced, by, among others, Frege, Wittgenstein, the Vienna Circle during its "formal mode of speech" years, Quine and Church. We have also seen that Tarski upheld a version of the same ineffability thesis as applied to his "colloquial language". I have argued elsewhere that the idea of the ineffability of semantics, in a somewhat generalized form, is the mainstay of the methodology of hermeneutical and deconstructivist approaches to philosophy. A great deal thus rests on the question of the definability of truth, especially on the question whether we can define truth for a realistic working language in that language itself. It is generally thought that all that there is unproblematical to logic (at least to the kind of logic a mathematician has occasion to use) is first-order logic. Everything else depends on a set theory, formulated as a first-order axiomatic theory. But set theory is not a part of logic, but a part of mathematics. Hence it cannot provide any absolute basis for the rest of mathematics, and is itself beset with all the problems of set existence.

In this book, I will show that this defeatist picture is wrong. Not surprisingly, a central part of this program concerns the ideas of truth and truth definition. It turns out, however, that in order to do so I will first have to revise some of our common ideas about truth and other semantical ideas, as well as about the foundations of logic.

Tarski's result seems to suggest – and even to establish – a negative answer to the problem of a realistic, philosophically interesting definability of truth. As such, Tarski's result is but a member of a family of apparently negative results which also include Gödel's incompleteness results and Lindström's (1969) theorem, according to which there cannot be a stronger logic than first-order logic (satisfying certain conditions) which has the same nice properties as first-order logic, principally compactness and the upward Löwenheim–Skolem property. Such negative results have dominated the thinking of philosophers of mathematics in recent decades. In my considered judgment, their importance has been vastly exaggerated.

A third function of logic in mathematics has cropped up in the course of my discussion above. It is the use of first-order logic as a medium of axiomatic set theory. This set theory is in turn supposed to serve as the universal framework of all mathematics.

This conception of set theory as the *lingua universalis* (or at least as a *lingua franca*) of mathematics is not universally accepted, and it is in any case riddled with difficulties. These problems will be discussed more fully later in this book, especially in Chapter 8. As a preview of what is to follow, the following problem areas are perhaps worth mentioning here:

(i) The current conception of set theory as axiomatic and deductive theory is quite different from the conception of set theory prevalent earlier. Mathematicians like Hilbert thought of set theory, not as a theory among many, but as a super-theory, a theory of all theories, which were thought of model-theoretically rather than deductively.

(ii) For reasons voiced so forcefully by Tarski, the conceptions of a universal theory and universal language for all mathematics of any sort is wrought with serious problems. For one thing, such a universal theory must inevitably be deductively incomplete.

(iii) Axiomatic set theory likewise is inevitably incomplete deductively. Because of this incompleteness, the necessary assumptions of set existence are extremely tricky.

However, this is not the end of the woes of axiomatic set theory. The full horror story is not revealed until we take a closer look at the different notions of completeness and incompleteness in Chapter 5 and then apply them to axiomatic set theory in Chapter 8.

Thus it is generally (but not universally) thought that the basic difficulties in the foundations of mathematics are problems of set existence, and more generally that the true core area of mathematics is set theory.

At this point it may be instructive to return to the relation of the first two functions of logic to each other. Obviously, the descriptive function is the basic one. Needless to say, one can try to study possible inferences by reference to their perceived appeal to us and systematize those that seem to be acceptable. The current euphemism for such an appeal to people's more or less educated prejudices is "intuition". Such a study of our "logical intuitions" nevertheless soon reaches a point where we need a firmer foundation for our system of logical inferences. The people's, and even the philosophical logician's, so-called intuitions have turned out to be fallible, and even mathematicians have not reached unanimity as to which principles of inference they should rely on. If an example is needed, then the checkered history of the axiom of choice serves as a case in point.

The right prescription for these inferential woes is to pay attention to the descriptive function of logic. This is what shows which putative logical inferences really preserve truth – and why. For instance, why is it that we can infer S_1 and S_2 from $(S_1 \mathbin{\&} S_2)$? Because of the meaning of "&" as the connective which combines sentences in such a way that both of them have to be true in order for the combination to be true.

More generally, the basis of all model theory is the relation of a sentence S to the set $M(S)$ of its models. A putative inference from S_1 to S_2 is then valid if and only if $M(S_1) \subseteq M(S_2)$.

Many variations are possible here, but they do not affect the main point. Logical inferences are based on the meaning of the symbols they involve, and purely logical inferences rely only on them.

I am painfully aware that distinctions of the kind I am making here have been challenged by certain philosophers. The bases of their theses are abstract views of meaning and the evidential status of meaning ascriptions. Here I am talking about the concrete issues in the foundations of logic and mathematics. And there the distinctions I am talking about have an unmistakable import.

This can be illustrated by pushing my line of thought further. From what I have said, it follows that a study of the inferences people are inclined to draw qualifies as a genuine logic only in so far as those inferences are based on the descriptive function of part of the logic in question. What this means in practice can be seen for instance from the so-called nonmonotonic logics. They may be magnificently interesting and important, but they are not logics in the sense just explained.

In the premises and conclusions of nonmonotonic inferences the logical constants clearly have their normal meaning. Hence the unusual principles of inference studied in these "logics" are not based on the same model-theoretical meaning of logical notions as in our ordinary logic. They are based on something else. And it is not very hard to see what this "something else" is. For instance, in what is known as circumscriptive inferences, the "something else" is the assumption that the premises in some sense supply all the information that is relevant to the subject matter. In some cases, this means that all the relevant individuals are mentioned in the premises or that their existence is implied by the premises (cf. Hintikka 1988c). Such assumptions are interesting, and they can be important both in theory and in practice, but they belong to the study of human communication, and not to logic. Aristotle would have dealt with them in the *Topics*, not in the *Analytics*.

This is not merely a terminological matter. It makes a difference to the way the "nonmonotonic inferences" are studied. For instance, inferences by circumscription ought to be studied by reference to their model-theoretical basis and not merely by postulating certain new patterns of inference. What we have in "nonmonotonic logics" is a clear instance of overemphasizing the deductive function of logic at the expense of its descriptive function.

In reality, the core area of logic is the one where all the valid inferences are based on the model-theoretical meaning of logical constants. Whether one wants to call what goes beyond this core area "logic" or not is a matter of intellectual taste. Someone else might all it heuristics or formalized psychology of reasoning. The important thing is to be clear about what one is doing.

Notes

[1] For this famous quip, see Blumenthal (1935, p. 403) and Toepell (1986, p. 42). A more pedestrian but at the same time perhaps more persuasive evidence for Hilbert's concern with the purely logical relations between the axioms and the theorems is forthcoming from the development of Hilbert's thinking (Toepell 1986). Of importance, a major role in Hilbert's development was played by questions as to which axioms are those that results like Pascal's Theorem and Desargue's Theorem depend or do not depend on.

[2] This question of the status of the axioms of continuity was at the time a matter of considerable interest. In the philosophy of physics, thinkers like Mach had flaunted the idea of a presuppositionless, purely descriptive phenomenological science typically describing the relevant phenomena by means of differential equations. Against them, Boltzmann pointed out vigorously that even the applicability of differential equations rested on nontrivial empirical assumptions concerning the phenomena, namely, on assumptions of continuity and differentiability (see, e.g., Boltzmann 1905). It would be interesting to know who influenced whom in this emphasis on the empirical content of continuity assumptions.

[3] Benacerraf (1973) in Benacerraf and Putnam (1983) attributes such a view to David Hilbert. In my view, this is a radical misinterpretation of Hilbert's ideas; see Hintikka (1996) and Chapter 9.

2

The Game of Logic

The pivotal role of truth definitions in the foundations of logic and mathematics prompts the question whether they can be freed from the severe limitations which Tarski's impossibility result apparently imposes on them – and whether they can be freed from other alleged or real defects that critics claim to have found in them.

One defect for which Tarski-type truth definitions are blamed is that of excessive abstractness. It has been alleged by, among others, *soi-disant* intuitionists and contructivists, that such definitions merely characterize a certain abstract relationship between sentences and facts. But such definitions leave unexplained, so this line of thought goes, as to what it is that makes this relation a truth relation. In particular, such abstract relations are unrelated to the activities by means of which we actually verify and falsify sentences of this or that language, whether a natural language or a formal (but interpreted) one. As Wittgenstein might have put it, each expression belongs to some language game which gives that expression its meaning. A specification of truth-conditions does not provide us with such a game, as Michael Dummett has doggedly argued time and again (see, e.g., Dummett 1978, 1991).

Criticisms like these have a good deal to say for themselves. There is much to be said for the fundamental Wittgensteinian idea that all meaning is mediated by certain complexes of rule-governed human activities which Wittgenstein called language games. Much of Wittgenstein's late philosophy is devoted to defending this fundamental vision against actual or potential – usually potential – criticisms, as is argued in Hintikka (1993b).

The kinds of criticisms I am talking about are often expressed in terms of a need to replace a truth-conditional semantics by a verificationist one. The philosophers stressing such a need nevertheless uniformly overlook the fact (pointed out in Hintikka 1987) that the constant between truth-conditional and verificationist semantics is not exclusive. For perfectly good truth-conditions can in principle be defined in terms of the very activities of verification and falsification. Indeed, it can be argued that such a synthesis is implicit in Wittgenstein's philosophy of language (cf. Hintikka and Hintikka 1986, Ch. 8). The deep idea in Wittgenstein is not that language can be used in a variety of ways, most of them nondescriptive; rather, Wittgenstein's point is that descriptive meaning itself has to be mediated by rule-governed human activities, that is, by language games. Moreover, the first language games Wittgenstein considered were games of verification and falsification. In such games, meaning can be both truth-conditional and verificationist, in that the truth-conditions themselves are, as it were, created and maintained by language games of verification and falsification.

Wittgenstein himself does not pay much systematic attention to such Janus-faced language games, and his self-designated followers have almost totally failed to appreciate them. In this chapter, I will nevertheless show that the idea of language games can be made a cornerstone of an extremely interesting logico-semantical theory. In doing so, I will also uncover an important additional ambiguity, this time an ambiguity affecting the notion of verification.

Furthermore, it will be shown – more by example than by argument – that the involvement of humanly playable language games does not make a concept of truth any the less objective or realistic.

These programmatic remarks, however, need to be put into practice. There is an obvious way of dealing with the difficulties about truth definitions. It is to confront the problem directly and to ask: What *are* the relevant language games, anyway, that constitute the notion of truth? How do we in fact verify and falsify sentences?

Let us take a simple example. How can (and must) you verify an existential sentence of the following form?

$$(\exists x)S[x] \tag{2.1}$$

where $S[x]$ is quantifier-free? The answer is obvious. In order to verify (2.1), one must find an individual, say b, such that

$$S[b] \tag{2.2}$$

is true. Here etymology serves to illustrate epistemology. In several languages, existence is expressed by a locution whose literal translation would be "one can find". For the quality of the pudding its proof may be in eating it, but when it comes to existence the proof of the pudding is in finding it.

Propositional connectives can be treated in the same manner almost a fortiori. For instance, if you have to verify a disjunction $(S_1 \lor S_2)$, what you have to do is no more and no less than to choose one of the disjuncts S_1 and S_2 and verify it.

But what about more complex cases? I will use a formal (but interpreted) first-order language L as an example. In order to speak of truth and falsity in connection with such a language, some model **M** of L (which you can think of either as "the actual world" or as a given fixed "possible world") must be specified in which truth or falsity of the sentences of L is being considered. The domain of individuals of **M** is called do(**M**). That L has been interpreted on **M** means that each atomic sentence (or identity) in the vocabulary of L plus a finite number of individual constants (names of members of do(**M**)) has a definite truth-value, true or false.

Consider now as an example a sentence of L of the form

$$(\forall x)(\exists y)S[x, y] \tag{2.3}$$

What is needed for me to be in a position to verify (2.3)? The answer is obvious. Clearly I must be able, given any value of x, say a, to find a value of y, say b, such that $S[a, b]$ is true. The only difference as compared with (2.1) is that now the individual to be looked for depends on the individual given to the verifier as the value a of the variable x.

What is needed for the purpose of making the finding of a suitable b a veritable test case of the truth of (2.3)? Clearly we have a test case on our hands if the value a of x is chosen in the most unfavorable way as far as the interests of the verifier are concerned. Descartes might have conceptualized this idea by letting the choice of a be made by a *malin genie*. It is nevertheless more useful to pick up a clue from John von Neumann rather than René Descartes and to think of that critical choice made by an imaginary opponent in a strategic game.

The natural way of generalizing and systematizing observations of the kind just made is therefore to define certain two-person games of verification and falsification. The two players may be called the initial *verifier* and the initial *falsifier*. I have called such games *semantical games* and an approach to the semantics of both formal and natural languages *game-theoretical semantics*, in short GTS.

The semantical game $G(S_0)$ associated with a sentence S_0 begins with S_0. At each stage of the game, the players are considering some sentence or other S_1. The entire game is played on some given model **M** of the underlying language.

On the basis of what has been said, the rules for semantical games are thoroughly unsurprising:

(R. ∨) $G(S_i \vee S_2)$ begins with the choice by the verifier of S_i (i = 1 or 2). The rest of the game is as in $G(S_i)$.

(R.&) $G(S_1 \,\&\, S_2)$ begins with the choice by the falsifier of S_i(i = 1 or 2). The rest of the game is as in $G(S_i)$.

(G.E) $G((\exists x)S[x])$ begins with the choice by the verifier of a member of do(**M**). If the name of this individual is b, the rest of the game is as in $G(S[b])$.

(G.A) $G((\forall x)S[x])$ likewise, except that falsifier makes the choice.

(R. ∼) $G(\sim S)$ is like $G(S)$, except that the roles of the two players (as defined by these rules) are interchanged.

(R.At)   If A is a true atomic sentence (or identity), the verifier wins $G(A)$ and the falsifier loses it. If A is a false atomic sentence (or identity), vice versa.

Since each application of one of the rules (R. ∨)–(R. ∼) eliminates one logical constant, any game $G(S)$ reaches in a finite number of moves a situation in which (R.At) applies – that is, a situation which shows which player wins.

It is to be noted that the name b mentioned in (R.E) and (R.A) need not belong to L. However, because of the finite length of any play of a semantical game, only a finite number of new names is needed in the language necessary to cope with any given play of a semantical game.

The rule (R.At) requires a special comment. It introduces an apparent circularity into my treatment in that it contains a reference to the truth or falsity of atomic sentences. However, as was pointed out above, the concept of truth can be applied to the relevant atomic sentences as soon as all the nonlogical constants of the given sentence have been interpreted on the given model **M** with respect to which the truth or falsity of S is being evaluated and on which $G(S)$ is being played. This interpretation is part and parcel of the definition of **M**. It is determined by the meanings of the nonlogical constants of *S*.

What (R.At) hence codifies is a kind of division of labor. The game-theoretical analysis of truth takes the meanings of primitive nonlogical constants of an interpreted first-order language for granted. This fixes the truth-values of the relevant atomic sentences, that is, of all sentences that can serve as endpoints of a semantical game. What my characterization does is to extend the notion of truth to all other sentences of the language in question.

The fact that I am thus restricting my task does not mean that I consider a further model-theoretical analysis of meanings unnecessary. My only reason for the restriction is that otherwise the scope of my enterprise would become unrealistically and unmanageably large.

It is important to realize, however, that what is taken for granted here is merely symbol meaning. The notion of sentence meaning is inextricably tied to the notion of truth. As one might say, a sentence means what it means by showing us what the world is like when the sentence is true. Thus the notion of truth is the be-all and end-all of sentence meaning in general.

To return to the rules of semantical games, it is important to realize that the notion of truth is in no way involved in the explicit formulation of the rules that govern the way moves are made in semantical games. Only on a heuristic level can it be said that their guiding idea is that *S* is true if and only if the applications of the game rules can always be chosen by the initial verifier so as to be truth-preserving. This heuristic idea leads also to the following game-theoretical truth definition for applied first-order languages:

(R.T)   *S* is true in **M** if and only if there exists a winning strategy for the initial verifier in the game G(*S*) when played on **M**.

Correspondingly, the falsity of a sentence can be defined

(R.F)   *S* is false in **M** if and only if there exists a winning strategy in G(*S*) for the initial falsifier.

There is an apparently small but in reality most consequential difference between the ideas represented here and those of constructivists like Dummett. They are not averse to using notions from strategic games in explaining their ideas but they give the game analogy a wrong, or perhaps a far too simplistic, turn. For instance Dummett writes (1978, p. 19):

The comparison between the notion of truth and that of winning a game still seems to me a good one.

But this specific analogy is a bad one. The interesting analogy is between the notion of truth and the existence of a winning strategy. In this respect, semantical games differ essentially from the "games" of formal proof. There the analogue to logical truth is winning in a single play of the game of proof-searching. All of this illustrates the subtleties – and the importance – of the apparently obvious truth definition (R.T).

These semantical games and the truth definition based on them can be extended in different directions. A similar approach can also be used in the semantics of natural languages. The treatment of first-order languages by means of semantical games is a paradigmatic example of what has been game-theoretical semantics (GTS). An extensive survey of GTS is presented in Hintikka and Sandu (1996).

This is not the occasion to advertise the merits of GTS. Its applications for natural languages speak for themselves. They are partially expounded in Hintikka and Kulas (1983, 1985) and in Hintikka and Sandu (1991). The present book can be considered another application and further development of the ideas of game-theoretical semantics. This further development is not prompted by a feeling of satisfaction with the existing theory, but by the questions and puzzles it gives rise to. At the same time, an examination of these open questions helps to put GTS itself into sharper focus.

The puzzles I am talking about are typically not difficulties for the development of the theory, but curious phenomena that suggest that further explanations are needed. In particular, my game-theoretical truth definition and the semantical games on which it is based require – and deserve – a number of further comments.

(i) The truth definition uses the notion of a winning strategy. Here the notion of strategy is used in its normal game-theoretical sense, which can be understood on the basis of the everyday sense of the word "strategy" but is stricter than that. In my sense, a strategy for a player is a rule that determines which move that player should make in any possible situation that can come up in the course of a play of that game.

This notion of strategy is the central concept in the mathematical theory of games. By using it, any game can be assumed to be represented in a normal form in which it consists simply of the choice of a strategy by each player. Together, these choices completely determine the course of a play of the game, including an answer to the question of who wins and who loses. A winning strategy for a player is one which results in that player winning no matter which strategy is chosen by the other player or players.

These notions are rather abstract, even though the starting-points of the abstraction are familiar and clear. This abstractness is a problem for a philosophical analyst, but it is also an opportunity for further development. I will avail myself of this opportunity in Chapter 10.

(ii) This kind of truth definition is not restricted to formal (but interpreted) first-order languages but can be extended to various other logical languages. It can also be extended to natural languages. Even though quantifiers (quantifier phrases) behave in certain respects differently in natural languages from the way they behave in the usual formal first-order languages, a treatment can be presented for them, too, in the same game-theoretical spirit. What is especially important here is that the very same truth definition applies there, too, without any changes. In other words, even though the game rules for particular moves are different, precisely the same characterization of truth and falsity can also be used in the semantics of natural languages.

(iii) Thus we have reached a semantical treatment of first-order languages and a characterization of truth which is in many ways a most satisfactory one. Its naturalness can be illustrated by telling evidence. The naturalness of the game-theoretical treatment of quantifiers is illustrated by the fact that it was put forward completely explicitly by C. S. Peirce (Hilpinen 1983) and that it has been spontaneously resorted to by logicians and mathematicians practically always when the usual Tarski-type truth definitions do not apply; and indeed sometimes when they do apply, as for instance in the Diophantine games of number theorists like Jones (1974).

Some of the reasons why Tarski-type truth definitions fail will be discussed in the later chapters of this book. Some others can be noted here. One of them is due to the fact that Tarski-type truth definitions start from the truth-conditions of the simplest (atomic) sentences and work their way recursively to the complex ones. This presupposes that there always are fixed starting-points for such a procedure; in other words, that the formulas of one's language are well-founded as set theorists would say. It is nevertheless possible to introduce, use, and to study languages which do not satisfy this requirement. Cases in point are game quantifier languages and more generally the infinitely deep languages first introduced in Hintikka and Rantala (1976). For such languages, it is impossible to give Tarski-type truth definitions. In contrast, game-theoretical characterizations of truth are perfectly possible. The only novelty is that some plays of a game can now be infinitely long. But, for a game theorist this is no obstacle to a definition of winning and losing. And once these notions are defined, the rest of one's GTS operates as of old.

More generally speaking, GTS is little more than a systematization of the mathematicians' time-honored ways of using quantifiers and of thinking of them. Careful mathematicians habitually use locutions like "given any value of x, one can find a value of y such that $\cdots$." A typical context of such talk is the so-called epsilon-delta definition of a notion like limit or derivative. An independent testimony might be more persuasive here than my own words. Speaking of the concept of limit, Ian Stewart writes in a recent book:

Finally $\cdots$ Karl Weierstrass sorted out the muddle in 1850 or thereabouts by taking the phrase 'as near as we please' seriously. How near *do* we please? He treated a variable, not as a quantity actively changing, but simply as a static symbol for any member of a set of possible values. (Stewart 1992, p. 105)

In other words, Weierstrass used quantifiers to analyze the concept of limit. But how did Weierstrass treat quantifiers? Stewart continues:

A function $f(x)$ approaches a limit L as x approaches a value a if, given any positive number ε, the difference $f(x) - L$ is less than ε whenever $x - a$ is less than some number δ *depending on* ε. It's like a game: 'You tell me how close you want $f(x)$ to be to L; then I'll tell you how close x has to be to a.' Player Epsilon says how near *he* pleases; then Delta is free to seek his own pleasure. If Delta always has a winning strategy, then $f(x)$ tends to the limit L. (Stewart 1992, pp. 105–106)

The only word in Stewart's account that I do not like is 'like' (in "*like a game*"), for what he describes is precisely the truth-condition for the $\varepsilon - \delta$ quantifier definition of limit in game-theoretical semantics.

However, the game-theoretical treatment of truth in interpreted first-order languages does not yet satisfy everything that philosophers might legitimately ask of it. First, what my treatment yields are *truth-conditions* for different first-order sentences. They are not united into a genuine *truth definition* or *truth predicate*. Such a definition must be formulated in a metalanguage in which we can speak of the syntax of the given first-order object language. Now we can discuss the syntax of a given first-order language in another first-order metalanguage, provided that the latter contains a modicum of elementary arithmetic, for instance by using the well-known technique of Gödel numbering. A truth definition will then have to consist in the definition of a number-theoretical predicate $T(x)$ which applies to the Gödel number of a sentence if and only if this sentence is true in the model under consideration. Such definitions for interpreted first-order object languages cannot be formulated in a first-order metalanguage. A fortiori, a truth definition for a given interpreted first-order language cannot be formulated in that language itself.

In these respects, my game-theoretical truth-conditions do not alone help us with the crucial problems as indicated in Chapter 1.

(iv) What the game-theoretical approach does tell us is what the truth-conditions of first-order sentences are like. These truth-conditions are formulated in terms of strategies of the two players. Now the notion of strategy is itself amenable to a logical analysis and to a formulation in logical terms. Suppose that a first-order sentence S is in a negation normal form (i.e., all negation signs prefixed to atomic formulas or identities). Since every first-order formula can be brought to this form by an effective procedure, this is not a restrictive assumption. Then a strategy for the initial verifier is defined by a finite set of functions (known as choice functions or Skolem functions) whose values tell which individual the verifier is to choose at each of his or her moves. These moves are in S connected with its existential quantifiers and disjunctions. The arguments of these functions are the individuals chosen by the falsifier up to that point in a play of the game $G(S)$. Choice functions are second-order entities, and their existence or nonexistence can be expressed by a second-order sentence. In this way, the game-theoretical truth-condition of S can be expressed by a second-order sentence S^*, which can be considered as a translation of S.

Speaking more explicitly, S^* can be obtained through the following steps:

(a) Let $(\exists x)$ be an existential quantifier occurring in S within the scope of the universal quantifiers $(\forall y_1), (\forall y_2), \ldots, (\forall y_k)$. Replace each occurrence of x bound to $(\exists x)$ by $f(y_1, y_2, \ldots, y_k)$, where f is a new function symbol, different for different existential quantifiers. Omit the quantifier $(\exists x)$.

These functions are usually called in logic the *Skolem functions* of S.

(b) Let $(S_1 \vee S_2)$ be a disjunction occurring within the scope of the universal quantifiers $(\forall y_1), (\forall y_2), \ldots, (\forall y_k)$. Replace the disjunction by

$$((S_1 \,\&\, (g(y_1, y_2, \ldots, y_k) = 0)) \vee (S_2 \,\&\, (g(y_1, y_2, \ldots, y_k) \neq 0))$$

$$(2.4)$$

where g is a new function symbol, different for different disjunctions and different from the functions f mentioned in (a).

In this work, I will extend the usual terminology somewhat and also call the g functions Skolem functions.

(c) Prefix the resulting formula by

$$(\exists f_1)(\exists f_2) \ldots (\exists g_1)(\exists g_2) \ldots$$

$$(2.5)$$

where $f_1, f_2, \ldots$ are all the functions introduced in (a) and $g_1, g_2, \ldots$ all the functions introduced in (b).

The result S^* will be called the *second-order translation of S*. It expresses the game-theoretical truth-condition of S. It states how the truth of S is connected with the semantical games of verification and falsification described earlier.

(v) It may nevertheless be questioned whether the concept of truth in general is really illuminated by the game-theoretical conditions. The job that they do is to specify what quantificational sentences *mean* by specifying their *truth-conditions*. The notion of truth is here a merely auxiliary one, it seems. In other words, the first-order semantical games seem to be language games for quantifiers, and not for the concept of truth. This is apparently in keeping with the nature of these games as games of seeking and finding. The conceptual connection between quantifiers and the activities of seeking and finding is easy to appreciate, but there does not seem to be any equally natural link between semantical games and the notion of truth in general. This can be thought of as being illustrated also by the impossibility of defining truth of quantificational sentences in those first-order languages which receive their meaning from my semantical games. One can suspect here, as Wittgenstein would have done, that the concept of truth can only receive a use – and *ergo* a meaning – in the context of certain other language games. Hence a great deal of further work needs to be done here, unless we are ready to acquiesce to all the different kinds of incompletenesses that prevail here. I shall take up these questions later in this work.

Another set of puzzles concerns the relation of GTS to constructivistic ideas. In a sense, I seem to have realized the constructivists' dream. I have shown the extremely close connection which obtains between the game-theoretically defined concept of truth and the activities (semantical games) by means of which the truth and falsity of our sentence is established. The lack of such a connection in earlier truth definitions has been the favorite target of constructivists. Now this objection is completely eliminated.

But, paradoxically, none of the consequences which constructivists have been arguing for follow from the game-theoretical truth definition. In fact, as far as first-order logic is concerned, the game-theoretical truth definition presented earlier in this chapter is equivalent to the usual Tarski-type truth definition, assuming the axiom of choice. Indeed, the second-order truth-condition of a given sentence S defined earlier in this chapter is equivalent to S, assuming the normal (standard) interpretation of second-order quantifiers. Are the constructivists' aims vacuous? Something strange is clearly going on here. I will return to this puzzle in Chapter 10.

One particular facet of constructivistic ideas turns out to be a teaser, too. Given the definitions of truth and falsity (R.T) and (R.F) presented earlier, there is in general no reason to believe that the law of excluded middle should hold. For according to (R.T) and (R.F) it holds for a given sentence S only if either the initial verifier or initial falsifier has a winning strategy in G(S). But we know from game theory that there are many two-person zero-sum games in which neither has a winning strategy. A simple example is a game in which each of the two players chooses a natural number, independently of the other's choice. The player with the larger number wins.

Games in which one of the two players has a winning strategy are said to be *determined*. The assumption that the one or the other player has a winning strategy is known as a *determinacy* (or *determinateness*) assumption. Some such assumptions can be extremely strong, as is known for instance from the different versions of the axiom of determinacy in set theory (cf. here Fenstad 1971).

Hence the law of excluded middle cannot in general be expected to hold if truth is defined game-theoretically. This should warm the heart of every true constructivist, for the law of excluded middle has long been their favorite target of criticism. Yet in that supposed core area of contemporary logic, ordinary first-order logic, *tertium non datur* does hold. An optimist might speak here of a fortunate coincidence, while a pessimist might be led to wonder whether first-order logic is really representative at all of the different intriguing things that GTS shows can happen in logic. I will return to this matter, too, later in Chapter 7. There, and elsewhere in the rest of this book, it will turn out that ordinary first-order logic is a fool's paradise in that it offers a poor and indeed misleading sample of the variety of things that can happen in logic in general.

Yet another Pandora's box of puzzling questions concerns the overall character of semantical games, especially as to what they are *not*. Earlier, I referred to them as activities of attempted verification and falsification. Yet this identification is by no means unproblematic. The very terms "verification" and "falsification" have to be handled with great care. For what kinds of activities do we usually think of as being involved in the verification and falsification of propositions? Any ordinary answer is likely to include at least the following two kinds of processes:

(a) logical (deductive) inferences
(b) different kinds of scientific inference, for instance, inductive inferences

Yet it is important to realize that semantical games are different from both these kinds of activities. How? Why? If semantical games do not codify the ways in which we in our actual epistemological practice verify and falsify sentences, then what light can they shed on a realistic notion of truth?

This rhetorical question can be answered in two parts. First, the activity of logically proving something is a language game of its own, with its own rules different from the rules of semantical games.

Second, and most important, this language game is parasitic on semantical games. Indeed, an attempt to prove that S_1 logically implies S_2 can be thought of as a frustrated attempt to construct a model ("possible world") in which S_1 is true but S_2 is not. If truth is understood game-theoretically, then this means constructing a model (world) **M** in which the initial verifier has a winning strategy in the game $G(S_1)$ played on **M** but not in the game $G(S_2)$ played likewise on **M**. For the purposes of ordinary first-order logic, the rules for such attempted model construction can be read from the rules for semantical games formulated above. The result is a complete set of *tobleau* rules for first-order logic. By turning them upside down we then obtain a somewhat more familiar looking set of sequent calculus rules for first-order logic.

Hence ordinary deductive logic does not present an alternative to semantical games for the process of verification and falsification. The rules of deductive logic are themselves parasitic on the rules for semantical games.

Somewhat similar points can be made of the processes that are usually said to be the ways of verifying and falsifying propositions – for instance scientific hypotheses. They, too, are, conceptually speaking, parastic on semantical games.

We have here a situation reminiscent of Meno's paradox, but without a paradoxical conclusion. In order to find out whether a proposition is true, one has to know what it means for it to be true. What this means is that the language games of actual scientific or other real-life verification depend on the semantical language games that are constitutive of a sentence's truth in the first place.

For instance, assume that I have to verify in actual scientific or everyday practice a functional dependence statement of the form

$$(\forall x)(\exists y)S[x, y] \tag{2.3}$$

(with x and y taking real-number values). In order to verify (2.3), I naturally do not undertake to play a game against an imaginary or real opponent, waiting for him or her or it to choose a real-number value of x for me to respond to. What I typically do is to try to find out what the function is that relates x to y in (2.1), for instance, by means of a controlled experiment with x as the controlled variable and y as the observed variable. Suppose that that experiment yields the function $g(x) = y$ as its result. This means that $g(x)$ is a Skolem function of (2.3), that is, that the following is true

$$(\forall x)S[x, g(x)] \qquad (2.6)$$

But this means that $g(x)$ is the strategy function (or a part of one) which enables me to win the semantical game connected with (2.3), that is, whose existence is the truth-condition of (2.3) according to the game-theoretical truth definition.

This can be generalized. The task of actually verifying a sentence S in the sense of getting to know its truth does not mean that the initial verifier wins a play of the game $G(S)$. Such a win may be due merely to good luck or to a failure of one's opponent to pursue an optimal strategy. Coming to know that S is true means finding a winning strategy for this game. The latter enterprise makes sense in principle only to someone who masters the game $G(S)$ whose strategies I am talking about. Hence the games of actual verification and falsification are based on semantical games, which are conceptually the more fundamental kind of game. The games of actual verification and falsification are secondary language games parasitic on the semantical games I have explained. Confusing the two kinds of games with each other is like confusing actual warfare in the trenches with the task of a general staff planning a campaign.

In sum, even though a real-life process of verification does not involve playing a semantical game, it aims at producing the very information – a winning strategy function – whose existence guarantees truth according to GTS.

These observations are instructive in several ways. They show, first of all, that the very concepts of verification and falsification are ambiguous. The terminology (and ideology) of language games serves us well in explaining this point. There are on the one hand those games which serve to define truth. I have shown that semantical games can play this role. But because of this very function, such games cannot be thought of as games of establishing (proving, verifying) a sentence whose truth-conditions are, so to speak, already known, without falling prey to Meno's paradox. Are such games to be called games of verification? Either answer can be defended, but the interesting point is that both answers are possible.

On the other hand there are games where the truth-conditions of the relevant sentences are taken for granted, and the game serves merely to enable the inquirer to come to know that these conditions are in fact satisfied. It is again a good question whether such games should be called games of verification and falsification. An edge is given to this question by the fact that such games are not merely games of verification and falsification. They serve a purpose that goes beyond that of characterizing the notions of truth and falsity. They serve the purpose of coming to *know* the truth of the sentences in question.

Perhaps both of these two kinds of games deserve to be related to the notion of truth. In the former, what is at issue is the very notion of truth, *veritas*, whereas in the latter we are dealing with some particular truth, *verum*.

What has been found strongly suggests that much of the recent talk of verification, falsification, assertibility conditions and so forth, is based on a confusion between the two related but importantly different kinds of language games which I have just distinguished from each other. The "real-life" or "real science" verification processes can in fact be subsumed under the approach to inquiry and reasoning in general which I have called the interrogative model of inquiry. In it, an inquirer reaches new information through a series of questions put to a suitable source of information ("oracle"), interspersed with logical inferences from answers previously received (plus initial premises). This model is really much more than a mere "model"; it is a new overall framework for epistemology. I have studied it in a number of publications, jointly with my associates (see, e.g., Hintikka 1985, 1988a). The interrogative model of inquiry can also be formulated as a game (language game). Prima facie, it does not look altogether different from the semantical games being examined here. Hence a confusion between the two is natural, albeit inexcusable. In reality, the two kinds of games are conceptually different from each other in extremely important ways. For one thing, interrogative games depend, as was indicated, on rules of logical inference. If they are changed, what can be verified by means of interrogative inquiry will also change. In fact, because of the epistemic element of interrogative inquiry, it may very well be argued that the usual deductive rules of first-order logic have to be changed in order to serve this purpose. If interrogative games were the logical home of the notion of truth, then changes in the rules of logical inference would affect the concept of truth. This would be in keeping with the thinking of constructivists like Dummett, who propose to discuss the "enigma" of truth by reference to rules of logical inference, and in fact advocate modifying the classical rules of first-order deductive inference. But once we realize that the language games which are constitutive of the notion of truth are semantical games, and not the games of interrogative inquiry, then the basis of this constructivistic line of thought vanishes. For semantical games are more basic than the "games" of formal logical proof. You may change the rules of deductive inferences by changing the rules of semantical games. Later, in Chapter 10, I will give concrete examples. But there is no way in evidence by which we could naturally do the converse. Moreover, the move-by-move rules of semantical games of the kind formulated above are so obvious that they offer scarcely any foothold for a critic to try to change them. Even the changes that will be considered experimentally in Chapter 10 will not affect these move-by-move rules.

One can say more here, however. If you seriously examine interrogative games, it will become overwhelmingly clear that the interrogative inquiry they model can only be understood and studied adequately as aiming at *knowledge*, and not merely at truth. Otherwise, there is no hope of understanding such crucial notions as the question–answer relationship, the criteria of conclusive answerhood, the presuppositions of different questions, and so forth. Most important, without acknowledging the epistemic element in interrogative inquiry we cannot cope with the use of such inquiry for the all-important purpose of answering questions, and not merely deriving given conclusions from given premises. Hence interrogative inquiry must be viewed as an essentially epistemic language game, aiming at knowledge and not merely at truth. And by the parity of cases, we must view the sundry processes which pass as verification and/or confirmation in ordinary usage and of which the interrogative model is a generalized model.

Thus we have to distinguish from each other three different kinds of language games. Their most important characteristics can be summed up in the accompanying table on p. 38.

In this table, the question mark indicates that winning a play of the game G(S) is not a hallmark of the truth of S. If it were, Meno's paradox would be applicable and semantical games could not be used to define truth because that notion would be needed to define and to understand semantical games. As it is, truth is defined indirectly as the existence of a winning strategy.

We might also speak of epistemic games as truth-seeking games and of semantical games as truth-constituting games.

GAME TYPE	Semantical game	Proof game	Interrogative game
WHAT WINNING SHOWS	?	logical truth	knowledge of truths
CRITERION OF WINNING	truth of the output sentence	closure of the game tableau	closure of the game tableau
OPERATIONAL-IZATION	seeking and finding	attempted counter-model construction	questioning plus logical inference

There are closer relationships between the three kinds of games. A proof "game" connected with S can be thought of as an attempted construction of a model in which S is not true, that is, in which there is no winning strategy for the initial verifier. An interrogative game is like a proof game except that answers to questions can be added as new premises. In spite of these connections, it is philosophically very important to distinguish them clearly from each other. Semantical games are outdoor games. They are played among the objects of the language one speaks, and they consist largely of the choices of the two players between different objects. In contrast, proof games are indoor games. They are played with pencil and paper, with chalk and chalkboard, or these days most likely with a computer.

Note also that none of the three types of games is a dialogical game in the literal sense of the word. The last two are "games against nature" in a game theorist's sense. For instance, all that the inquirer's opponent does in an interrogative game is occasionally to answer questions.

Once you develop a theory of interrogative games of inquiry, you will also see that constructivists like Dummett are making further unwarranted and distortive assumptions concerning the character of these "games against nature". Even though I have not found a "smoking gun" statement in his writings, a number of things Dummett says (e.g., 1978, p. 227) makes sense to me only if the empirical input into the process of verification and falsification consists entirely of observations whose expressions in language are assumed to be atomic propositions. This fits very well into Dummett's paradigm case of arithmetic, where the directly verifiable propositions are numerical ones. In general, this assumption amounts to claiming that the only answers one can hope to obtain in interrogative games are particular truths. I have called this assumption the atomistic postulate. A closer examination shows, however, that not only is this assumption unwarranted, but that it has distorted philosophers' entire view of scientific inquiry. In this respect, too, Dummett's ideas of what the processes ("language games") are like which we actually use to verify and falsify our sentences are inaccurate.

Of course Dummett and his ilk have not enjoyed the privilege of considering my interrogative model in so many words. What they have had in mind is some, more or less, unarticulated idea of how we in practice verify and falsify scientific and everyday propositions. But the interrogative model is calculated to capture all such activities. Indeed, the only main assumption I have to make for the applicability of the interrogative approach is that all new information imported into the argument can be thought of as being obtained as an answer to an implicit (or explicit) question addressed to a suitable source of information. For this to be possible, little more is required than that the inquirer is aware where he or she gets his or her information from. Hence the interrogative model applies extremely widely to all sorts of different processes of knowledge-seeking and verification, presumably including what constructivists have in mind. Hence they are not necessarily wrong about our actual processes of inquiry. Their mistake is to assume that these processes are constitutive of the notion of truth. They have missed the point of Meno's paradox. To seek truth, one has to know what truth is.

Hence the usual scientific and/or everyday procedures of verification cannot serve to define truth. This suggests that semantical games might ultimately be used to define an actual truth predicate. Whether my arguments show that semantical games can do so depends on what will be found about such notions as Skolem function. I will return to such questions in Chapter 6.

The distinction between semantical games on the one hand and "games" of formal proof as well as interrogative games on the other hand is thus, in any case, a crucial one. At the same time, the distinction puts the nature of semantical games into sharper profile.

What consequences or suggestions does the game-theoretical approach (and its success) to logic yield? The rest of this book is but one long argument calculated to answer this question. One specific suggestion might nevertheless be registered here. In order to see what it is, we might consider what GTS says of truth of sentences of the form

$$(\forall x)(\exists y)S[x, y] \qquad (2.3)$$

where $S[x, y]$ is (for simplicity only) assumed not to contain quantifiers. According to GTS, (2.3) is true if and only if there exists a winning strategy for the initial verifier in the correlated game. Such a strategy will tell among other things how to choose the value of y depending on the value of x. This part of the verifier's strategy can therefore be codified in a function $f(x) = y$. For such a function to be part of a winning strategy means that the following sentence is true:

$$(\exists f)(\forall x)S[x, f(x)] \qquad (2.7)$$

Conversely, (2.7) is obviouly true if (2.3) is.

Hence (2.3) and (2.7) must be true simultaneously. But the equivalence of (2.3) and (2.7) is but a variant of the axiom of choice. Hence GTS resoundingly vindicates this controversial axiom. Moreover, the reasons for the validity of the axiom of choice are purely logical. The axiom is but a corollary of the game-theoretical definition of truth. Hence GTS realizes Hilbert's (1923, p. 157) bold conjecture, not only vindicating the axiom of choice but showing that it is as unexceptionable a truth as $2 + 2 = 4$.

In one sense, it is an understatement to say that the game-theoretical viewpoint vindicates the axiom of choice because, by the same token, it vindicates any equivalence between any old first-order sentence and its second-order "translation" (truth-condition). Some such equivalences are known to be stronger than the plain-vanilla equivalences between (2.3) and (2.7), where $S[x, y]$ does not contain quantifiers or disjunctions (in its negation normal form). For instance, such simple equivalences cannot enforce the existence of nonrecursive functions while some of the more complex equivalences can (see Kreisel 1953; Mostowski 1955). Hence we apparently have here a hierarchy of stronger and stronger assumptions available to mathematicians. However, a separate investigation is still needed to see whether the more complex equivalences are actually stronger assumptions than the ordinary axiom of choice in a set-theoretical context. What has been seen in any case shows that the axiom of choice cannot be discussed in isolation. It is a member of a series of assumptions that can all be justified in the same way. Moreover, its justification or rejection is tied to more general questions of the nature of truth, especially to questions concerning the game-theoretical analysis of truth. In so far as I can vindicate the general ideas of GTS, I can to the same extent vindicate the axiom of choice. In Chapter 10 it will turn out that even certain interesting constructivistic changes in my game-theoretical approach fail to provide any reasons for rejecting the axiom of choice.

Gödel has advocated a cultivation of our mathematical intuition in ways that would eventually enable us to see the truth or falsity of different set-theoretical assumptions. In the spirit of Gödel's idea, it may be said that game-theoretical semantics provides us with the intuitions that are needed to vindicate the axiom of choice plus a number of related assumptions. I believe that this is a fair way of describing the situation. Moreover, the intuitions in question have been seen to be solidly grounded on the ways in which we actually deal with the concept of truth.

One can add a slightly different twist to the axiom of choice by recalling that (2.7) is but a restatement of the truth of (2.3) from a game-theoretical viewpoint. Hence the equivalence of (2.7) and (2.3) is for a game-theoretical semanticist an instance of Tarski's so-called T-schema. This schema is usually said to be of the form

$$\delta \text{ is true} \leftrightarrow S$$

where δ is a quote or a description of S. More generally, we can speak of an instance of the T-schema in our extended sense whenever an equivalence is set up between a sentence and its truth-condition. The equivalence between (2.7) and (2.3) is a case in point. Hence the axiom of choice is from the game-theoretical viewpoint closely related to the T-schema, a relationship which illustrates the intuitiveness of the axiom of choice in the framework of GTS.

The same intuitiveness is likewise associated with the stronger forms of the axiom of choice obtained by setting up equivalences between more complex first-order sentences and their respective game-theoretical truth-conditions.

A skeptical reader might very well remain unconvinced. But even the most dogmatic skeptic – or should I say, the most skeptical skeptic? – now knows what he or she must focus their critical scrutiny on. All evidence for GTS is *ipso facto* evidence for the axiom of choice. Any interesting criticism of the axiom of choice must be based on an examination of the basic ideas of GTS, and in particular of the game-theoretical definition of truth as the existence of a winning strategy for the initial verifier in a semantical game. Indeed, I will myself subject this definition to a closer examination in Chapter 10. Hence the reader will have to remain in suspense, till the last episode of the story of the perils and triumphs of the axiom of choice.

Furthermore, it is not only the concepts of verification and falsification (as well as the contrast between truth-conditional and verificationist semantics) that are shown by GTS to need further attention and further distinctions. It is likewise seen that the received trichotomy of syntax – semantics – pragmatics has to be reconsidered, for where does game-theoretical semantics belong in this classification? It deals with some of the basic semantical relations between language and the world, and hence should belong to semantics. At the same time, it deals with the uses of language in certain language games. Hence Charles Morris (1938), who is primarily responsible for the unfortunate syntax – semantics – pragmatics trichotomy, would presumably relegate it to what Yehoshua Bar-Hillel used to call the pragmatic wastepaper basket.

The right conclusion from my case study is that the traditional distinction between semantics and pragmatics involves two separate fallacies. First, crucial semantical relations like that of truth can exist (and perhaps must exist) only in the form of certain rule-governed human activities à la Wittgensteinian language games. Being activities of language use, they would have to be pigeonholed as belonging to pragmatics. In spite of this, their applicability to the question of the truth of a given sentence in some particular world does not depend on the human agents who implement these games. Their applicability depends only on the structure of the "world" (model) in question.

Hence the contrast between semantics and pragmatics must be defined in some other way, lest we end up saying that semantics is a part of pragmatics. Personally I would not mind such a conclusion very much, as long as the other fallacy muddling these two notions is avoided. This fallacy consists in thinking that all theories of language must inevitably involve the language users, and are therefore part and parcel of the psychology and sociology of language. This view is as fallacious as the claim would be that the study of syntax is a part of graphology (or in the present day and age, of the technology of computer displays and computer printouts). In both cases, we can (and do) abstract from the idiosyncracies of the persons (and computers) in question and concentrate on the general rules that govern writing or language use, respectively. If a dramatization of this fact is needed, then the reader is invited to think of how automata could be programmed to play semantical games as they perfectly well could be programmed to do. The hardware of such robots would be irrelevant to the semantical and logical situation – only the software would matter.

A further important contribution which the game-theoretical approach can make to the philosophical clarification of logic is an account of how the bound variables of quantification really work. What does such a variable "mean"? Such questions appear easy to answer only as long as you do not raise them in so many words. It is far from clear how a sentence in (say) a first-order formal notation is in fact processed psycholinguistically, in spite of the frequent practice of linguists to use in effect first-order logic as the medium of semantical representation, or as the medium of representing logical form. One's first idea is to think of bound variables as some sort of referring terms. But even though are similarities between the variables of quantification and ordinary singular terms, there are also differences.

Many of the same puzzles which arise in connection with the formal variables of quantification also arise in connection with the quantifier phrases of natural languages. This is not surprising, for one of the main uses of formal quantification theory is supposed to be to serve as the framework of semantical representation into which the quantifier sentences of natural languages can be translated (cf. Russell 1905).

For instance, two occurrences of the same singular term refer to the same individual. But two occurrences of the same quantifier phrase (or the same variable) need not do so. This can be illustrated by the contrast between the following sentences:

John admires John. (2.8)

Everybody admires everybody. (2.9)

If you want to find a sentence which is semantically parallel to (2.8), it is not the syntactic analogue (2.9) to (2.8) but rather

Everybody admires himself. (2.10)

In game-theoretical semantics, bound variables are placeholders for the names of the individuals which the players choose during a play of a semantical game. They are thus relative to such a play. They can be said to stand for particular individuals within the context of one particular play, but not absolutely. This helps to explain both the similarities and the dissimilarities between quantificational variables and ordinary singular terms. Further details are found in Hintikka and Kulas (1985).

For instance, each occurrence of a quantifier phrase has to be replaced by the name of an individual independently of others, even if the two phrases are formally identical. This explains the semantical difference between such sentences as (2.8) and (2.9).

In general, the game-theoretical approach facilitates a theoretically satisfying account of the behavior of the variables of quantification in formal languages and of quantifier phrases in natural languages, including several aspects of anaphora in natural languages (see here, e.g., Hintikka 1976a and Hintikka and Kulas 1985). This clarification of the basic notations of the logic of quantification by means of GTS is a natural continuation of the analysis of basic mathematical concepts by means of quantifiers carried out by the likes of Weierstrass. In this respect, the quote from Ian Stewart earlier in this chapter is a most instructive one.

The possibility of a game-theoretical concept of truth which accords with our natural concept of truth, together with the distinction between semantical (truth-conditioning) and interrogative (truth-seeking) games also has profound philosophical repercussions. For one thing, it shows what is true and what is false in pragmatist conceptions of truth. What is true is that to speak of truth is not to speak of any independently existing correspondence relations between language and the world. There are no such relations. Or, as Wittgenstein once put it, the correspondence between language and the world is established only by the use of our language – that is, by semantical games. Truth is literally constituted by certain human rule-governed activities.

What is false in pragmatist ideas about truth is the claim that the relevant activities are the activities by means of which we typically find out what is true – that is to say, verify, falsify, confirm, disconfirm and so forth, our propositions. This claim is based on overlooking the all-important distinction between truth-establishing games (that is, semantical games) and truth-seeking games (that is, interrogative or perhaps other epistemic games). For all their formal similarities and partial mutual determinations, these two kinds of language games are fundamentally different from each other philosophically and should be kept strictly apart. And when they are distinguished from each other and their true nature recognized, then the pragmatist claim is seen to be false. Our actual truth-seeking practices, whether or not they are relative to a historical era, epistemic or scientific community, social class or gender, are not constitutive of our normal concept of truth – that is, of *the* concept of truth.

3

Frege's Fallacy Foiled:
Independence-Friendly Logic

Several of the problems uncovered in the first two chapters above can be traced to a common source. This source is a somewhat surprising one. If there is an almost universally accepted dogma in contemporary logical theory and analytic philosophy, it is the status of the usual first-order logic as the unproblematic core area of logic. "If I don't understand it, I do not understand anything," fellow philosopher once said to me in discussing first-order logic. And when on another occasion I expressed reservations concerning the claims of the received first-order logic to be the true *Sprachlogik*, the logic of our natural language, a well-known philosopher's response was: "Nothing is sacred in philosophy any longer." He might have added, "Or in linguistics, either," with a side glance at the role of the usual quantificational (first-order) notation as the preferred medium of representing logical form (LF) in Chomsky's linguistic theorizing.

Yet the received first-order logic, as formulated by the likes of Frege, Russell and Whitehead, or Hilbert and Ackermann, involves an important restrictive assumption which is largely unmotivated – or perhaps rather motivated by wrong reasons. This assumption is the hidden source of the problems registered above. For simplicity, I will refer to it as Frege's fallacy or mistake. Though I will not discuss the historical appropriateness or inappropriateness of that label here, it is of some interest to note that Peirce would have found it much easier to avoid this mistake – if he committed it in the first place.

In order to diagnose Frege's mistake, we have to go back to the basic ideas of first-order logic. This part of logic is also known as quantification theory. It is in fact clear that the notions of existential and universal quantification play a crucial role in serving both the deductive and the descriptive function of logic, so much so that the birth of contemporary logic is usually identified with the roughly simultaneous discovery of the notion of quantifier by Frege and Peirce.

But to say that first-order logic is the logic of quantifiers is to utter a half-truth. If you consider the two quantifiers in isolation from each other you are not going to obtain an interesting logic – that is to say, a logic which could serve the purposes of a nontrivial language. All we would get is some mild generalization of syllogistic logic, for instance, monadic predicate logic.

No, the real source of the expressive power of first-order logic lies not in the notion of quantifier per se, but in the idea of a *dependent quantifier*. If you want to give a simple but representative example of a quantificational sentence, it will not be a syllogistic premise with just one universal or existential quantifier, but a sentence where an existential quantifier depends on a universal one, such as

$$(\forall x)(\exists y)S[x, y] \tag{3.1}$$

Without such dependent quantifiers, first-order languages would be pitifully weak in expressive power. For instance, without them we cannot express any functional dependencies in a purely quantificational language. The fact that existential quantification can be hidden under a functional notation does not change the situation. Such hidden quantifiers are forced out of the closet when function symbols are eliminated in favor of predicate symbols.

Hence it is no exaggeration to say that to understand first-order logic is to understand the notion of a dependent quantifier. If my learned friend just quoted is right, if you do not understand the notion of a dependent quantifier you do not understand anything.

One important service which game-theoretical semantics performs is to explain the nature of the dependence in question. In game-theoretical terms, this dependence is *informational dependence*. It means that when a player of a semantical game makes a move prompted by a dependent quantifier, then he or she knows what the earlier move is like that was prompted by the quantifier it depends on. In the technical jargon of game theory, the earlier move is in the information set of the other (later) move. For instance, in playing a game with (3.1), when the verifier chooses a value for y, he or she knows which individual was chosen by the falsifier as a value of x and can use this information in making his or her choice of the value of y.

This game-theoretical explanation of the nature of dependent quantifiers is no mean feat. At least it is superior to such meaner explanations of the nature of quantifiers as Frege's suggestion to construe first-order quantifiers as second-order predicates (predicates of one-place predicates), which serve to tell whether the given predicate is empty or nonempty, exceptionless or exception-admitting. Such explanations overlook the fact that quantifiers can syntactically speaking be applied to complex or simple predicates with more than one argument-place. This oversight in effect means disregarding the possibility of dependent quantifiers, and hence disregarding the true source of the expressive strength of first-order languages. It is not only the case that there are certain arcane phenomena in the semantics of natural language quantifiers that are not done full justice to by considering quantifiers as higher-order predicates. By so doing one just cannot bring out fully the most important features of the behavior of quantifiers, as I will argue later in this paper.

In the notation of the usual first-order logic, quantifier dependence is indicated by parentheses. Each quantifier is assigned a pair of parentheses indicating its scope. A quantifier occurring within the scope of another is dependent on it. This is not the only way of indicating dependence, however. For instance, in the gametheoretical second-order translations of first-order sentences the dependence of an existential quantifier on universal ones is shown by the selection of arguments of the Skolem function that replaces it. For instance in

$$(\forall x)(\exists y)(\forall z)(\exists u)S[x, y, z, u] \tag{3.2}$$

the dependencies are indicated by the tacit parentheses delineating the scopes of the different quantifiers. In the second-order translation of (3.2), to wit,

$$(\exists f)(\exists g)(\forall x)(\forall z)S[x, f(x), z, g(x, z)] \tag{3.3}$$

the dependencies are indicated by argument selection rather than parentheses. In general, the variable x bound to a given existential quantifier $(\exists x)$ is replaced in the second-order translation by a functional term $f(z_1, z_2, \ldots)$, precisely when the universal quantifiers within whose scope $(\exists x)$ occurs are $(\forall z_1), (\forall z_2), \ldots$ (I am once again assuming that the sentence in question is in the negation normal form.)

In contrast,

$$(\forall x)(\forall z)(\exists y)(\exists u)S[x, y, z, u] \qquad (3.4)$$

is equivalent to

$$(\exists f)(\exists g)(\forall x)(\forall z)S[x, f(x, z), z, g(x, z)] \qquad (3.5)$$

where the new argument z in $f(x, z)$ shows that $(\exists y)$ depends in (3.4) also on $(\forall z)$ and not only on $(\forall x)$, as in (3.2).

But the two methods of indicating dependence and independence are not equivalent. The Frege–Russell use of parentheses is more restrictive than the use of Skolem functions. For instance, consider the following second-order sentence which is on a par with (3.3) and (3.5):

$$(\exists f)(\exists g)(\forall x)(\forall z)S[x, f(x), z, g(z)] \qquad (3.6)$$

From the game-theoretical viewpoint, (3.6) is in perfect order. It expresses the existence of a winning strategy in a semantical game that is easily defined and implemented. The move-by-move rules of this game are the same as in the GTS of ordinary first-order logic. In this game, the initial falsifier chooses two individuals b, d to serve as the respective values of x and z. Then the initial verifier chooses a value of y, say c, knowing only the value of x; and chooses a value of u, say e, knowing only the value of z, in both cases with an eye to winning the game with $S[b, c, d, e]$. Admittedly, in order to implement such a game in a form which involves human players, we must think of the initial verifier as a team of at least two human beings. This is no hindrance to a game-theoretical logician, however. All that it means is that my semantical games can occasionally look more like bridge, where each "player" in the technical jargon of game theory is a pair of human beings possessing somewhat different information, than chess, where the game-theoretical usage of "player" coincides with the colloquial one. (More generally speaking, a game theorist's "players" can humanly speaking be teams of several men or women.)

But although there is a well-defined semantical game that corresponds to (3.6) and lends it a well-defined truth-condition, there is no formula of the traditional first-order logic which is equivalent to (3.6). This can be seen by trying to order linearly the quantifiers $(\forall x)$,

$(\forall z)$, $(\exists y)$, $(\exists u)$. If we indicate the order by $>$ we can argue as follows: Since $y = f(x)$ depends on x but not on z, we must have

$$(\forall x) > (\forall y) > (\forall z) \qquad (3.7)$$

Since $u = g(z)$ depends on z but not on x, we must have

$$(\forall z) > (\exists u) > (\forall x) \qquad (3.8)$$

But (3.7) and (3.8) are incompatible, thus proving my point.

If we devise a notation which allows partial rather than linear ordering, then we can easily formulate equivalents to (3.6). The following two-dimensional formula fits the bill:

$$\left.\begin{array}{l}(\forall x)(\exists y) \\ (\forall z)(\exists u)\end{array}\right\} S[x, y, z, u] \qquad (3.9)$$

It will be shown further on, however, that partially ordered quantifier structures are not the most natural way of capturing the phenomenon of informational independence.

Thus, we have arrived at a diagnosis of Frege's mistake. This diagnosis is virtually forced on us by the game-theoretical viewpoint. The first question any game theorist worth his or her utility matrix will ask about semantical games is: Are they games with perfect information or not? Looked upon from our anachronistic game-theoretical perspective, Frege in effect answered: Logic is a game with perfect information. In spite of its prima facie plausibility, this answer is nevertheless *sub specie logicae* not only arbitrary and restrictive but wrong, in that in cuts off from the purview of logical methods a large class of actual uses of our logical concepts. The fact that those uses are descriptive rather than deductive ought not to have deterred Frege, either. The crucial fact here is that the ideas Frege was overlooking are implicitly an integral part and parcel of a logic which everybody has practised and which almost everybody thinks of as the unproblematic core area of logic, namely, the logic of quantifiers.

Hence what is to be done here is clear. We have to extend our familiar traditional first-order logic into a stronger logic which allows for information independence where the received Frege–Russell notation forbids it. The result is a logic which will be called *independence-friendly* (IF) *first-order logic*. It is fully as basic as ordinary first-order logic. It is our true elementary logic. Everything you need to understand it, you already need to understand traditional Frege–Russell first-order logic. To quote my own earlier exposition, IF first-order logic is a true Mafia logic: It is a logic you cannot refuse to understand.

IF first-order logic can be implemented in several different ways, some of which will be discussed further on. The formulation used here involves one new item of notation, for which I will use the slash "/". It will be used to exempt a quantifier or a connective (or, for that matter, any logically active ingredient of a sentence) from the scope of another one. Thus, for instance

$$(\forall x)(\exists y/\forall x)S[x, y] \tag{3.10}$$

is like (3.1) except that the choice by the verifier of a value of y must be made in ignorance of the value of x. In this case, the former choice might as well be made before the latter. In other words, (3.10) is logically equivalent to

$$(\exists y)(\forall x)S[x, y] \tag{3.11}$$

In this particular case, the slash notation does not yield anything that we could not express without it. But in other cases, the independence-friendly notation enables us to express what cannot be expressed in the old notation. For instance, (3.6) can now be expressed on the first-order level as

$$(\forall x)(\forall z)(\exists y/\forall z)(\exists u/\forall x)S[x, y, z, u] \tag{3.12}$$

It is seen here how the slash notation serves to omit potential arguments of the Skolem (strategy) functions associated with a first-order sentence.

The quantifier structure exemplified by (3.12) (or, equivalently, by (3.9)) is known as the Henkin quantifier.

The slash notation, unlike the branching notation, can (and must) also be applied naturally to propositional connectives. For instance, in

$$(\forall x)(S_1[x](\vee /\forall x)S_2[x]) \tag{3.13}$$

the choice between $S_1[x]$ and $S_2[x]$ must be made independently of the value of x. Hence (3.13) is equivalent to

$$(\forall x)S_1[x] \vee (\forall x)S_2[x] \tag{3.14}$$

But in other cases the slash notation is uneliminable (on the first-order level) even when applied to propositional connectives. An example is

$$(\forall x)(\forall z)(\exists y/\forall z)\,(S_1[x,y,z] \quad (\lor/\forall x)\,S_2[x,y,z]) \tag{3.15}$$

In a second-order translation, (3.15) becomes

$$(\exists f)(\exists g)(\forall x)(\forall z)((S_1[x,f(x),z]\ \&\ (g(z)=0))$$
$$\lor (S_2[x,f(x),z]\ \&\ (g(z)\neq 0))) \tag{3.16}$$

which shows the parallelism between (3.15) and (3.12).

It turns out that for the purposes of IF first-order we do not even have to deal with all the different possible types of independence. It also suffices to explain the slash notation for formulas in negation normal form – that is, in a form in which all negation-signs are prefixed to atomic formulas or identities. Rules for transforming formulas into negation normal form and out of it are the same in IF first-order logic as in its traditional variant.

Thus we can explain the formalism of IF first-order logic as follows:

Let S_0 be a formula of ordinary first-order logic in negation normal form. A formula of IF first-order logic is obtained by any finite number of the following steps:

(a) If $(\exists y)S_1[y]$ occurs in S_0 within the scope of a number of universal quantifiers which include $(\forall x_1),(\forall x_2)\ldots$, then it may be replaced by

$$(\exists y/\forall x_1,\forall x_2,\ldots)S_1[y] \tag{3.17}$$

(b) If $(S_1 \lor S_2)$ occurs in S_0 within the scope of a number of universal quantifiers which include $(\forall x_1),(\forall x_2)\ldots$, then it may be replaced by

$$(S_1(\lor/\forall x_1,\forall x_2,\ldots)S_2) \tag{3.18}$$

Languages using IF first-order logic are called IF first-order languages.

IF first-order languages will be my main conceptual tool in the rest of this work. They deserve (and need) a number of further explanations. Earlier discussions of independence-friendly first-order logic are found in Sandu (1993) and Hintikka (1995b).

(i) A few remarks are in order concerning the formation rules for IF first-order logic. They will turn out to be more interesting than what might first appear.

In the form envisaged in (a) and (b), those formation rules for IF first-order languages that introduce slashes are of an unusual form. They take an already formed sentence (closed formula) and, instead of using it as a building-block for more complex expressions, modify certain subformulas of the expression in question. Moreover, these changes inside the given expression are context-sensitive. What changes are admissible depends on the context of the subformula in question. For instance, whether $(\exists y)$ can be replaced (*salva* well-formedness) by $(\exists y/\forall x)$ depends on whether $(\forall x)$ occurs somewhere in the larger formula with $(\exists y)$ within its scope.

I could eliminate this defect (if it is a defect) of the formation rules for IF logic by using a slightly different notation. In it, the independence of $(\exists y)$ from $(\forall x)$ (where the former occurs in the scope of the latter) is not expressed by replacing $(\exists y)$ by $(\exists y/\forall x)$ but by replacing $(\forall x)$ by $(\forall x//\exists y)$. It is obvious how this notation can be used in all the different cases.

For instance, (3.10) can be written in the alternative notation as

$$(\forall x//\exists y)(\exists y)S[x, y] \qquad (3.19)$$

while (3.12) would become

$$(\forall x//\exists u)(\forall z//\exists y)(\exists y)(\exists u)S[x, y, z, u] \qquad (3.20)$$

and (3.14) would become

$$(\forall x//\vee)(S_1[x] \vee S_2[x]) \qquad (3.21)$$

The double-slash notation is in fact somewhat more natural than the single-slash one. Unfortunately, I did not realize this fact when I chose my notation, and at least for the purposes of this book it is too late to change now. This does not matter very much, however, for the interesting thing here is a comparison between the two notations.

The naturalness of the double-slash notation is shown by the fact that the formation rules for IF first-order languages would be context-independent. Over and above this advantage, the double-slash notation has other advantages. For instance, it would enable us to formulate various simple logical laws. For instance, the equivalence of (3.13) and (3.14) will then be an instance of a distribution law for $(\forall x//\vee)$:

$$(\forall x//\vee)(S_1[x] \vee S_2[x]) \qquad (3.21)$$

will be equivalent to

$$(\forall x)S_1[x] \lor (\forall x)S_2[x] \qquad\qquad (3.22)$$

Even though the difference between / and // is in some sense only a matter of notation, it will nevertheless turn out in Chapter 6 that it illustrates certain important methodological problems.

(ii) Frege's fallacy is connected with an impotant unclarity concerning the notion of scope. This notion is used widely both in logic and in linguistics, and it is normally taken for granted. When it comes to first-order languages (and to their counterparts in natural languages), the very notion of scope nevertheless involves certain serious problems. They are once again brought out into the open by the game-theoretical approach. It shows at once that the traditional notion of scope has two entirely different functions when it is applied to quantifiers.

First, it is used to indicate the order in which the game rules are applied to a given formula. In other words, the nesting of scopes serves to indicate logical priority. This notion of scope will be called priority scope. In the received notation of first-order logic, priority scope is indicated by the inclusion of the scope of a quantifier (or a connective) in the scope of another. The latter will then have the right of way when it comes to game rule application. In order to serve this purpose, the scopes of quantifiers and of connectives must be partially ordered – that is, they must be nested. Furthermore, only the relative order of the scopes matters and not their actual extent in a sentence or discourse.

But the scopes of quantifiers are also supposed to indicate something else. When the scope of a quantifier (Qx) is indicated by a pair of parentheses, then the segment of the sentence within those parentheses is supposed to be where the variable x is bound to the quantifier in question. These variables are often compared to anaphoric pronouns. The scope of a quantifier is then the part of a sentence where such a pronoun can have the quantifier in question as its head. This kind of scope will be called the binding scope. The distinction between priority scope and binding scope is related to the difference between Chomsky's (1981) notions of government and binding, and could be used to elucidate what Chomsky is trying to get at in his theory.

Once the distinction is made, then it is seen that there is no reason why priority scope and binding scope should always go together.

For instance, there is no reason why the scopes of two quantifiers (in the purely formal sense of scope in logic) could be only partially overlapping, even though the Frege–Russell notation requires that they be either nonoverlapping or else nested (one of them contained in the other). It can even be argued that the notion of binding scope is not an unanalyzable one in natural languages. Furthermore, the binding scope of a quantifier need not be a continuous segment of a sentence or a discourse.

Once the situation is thought of in these terms, then it is seen that the usual notion of scope as it is being actually employed is a royal mess. This mess assumes a somewhat different complexion for logicians and linguists. For logicians running together the two senses of scope has been an unnecessary obstacle to their theorizing, but for linguists it has actually harmed their theories, in that it has led them to overlook certain telling phenomena of our *Sprachlogik*.

Frege's mistake can thus be seen as being a result of conflating the two uses of parentheses. The most natural way of using them for the purpose of indicating logical priority is to require that the scopes of quantifiers are nested. But to do so precludes a merely partial overlapping of quantifier scopes, even though such a violation of the nesting requirement is a perfectly natural thing from the vantage point of binding scope.

If Frege's needless restrictions on the use of scope indicators are removed, we then obtain a first-order language that is as strong as IF first-order language. If quantifiers are indexed together with the corresponding parentheses, we could then for instance express (3.12) as follows:

$$(\forall x)_{11}((\exists y)(\forall z)_2)_{12}((\exists u)_1(S[x,y,z,u])_1)_2 \qquad (3.23)$$

Here the scope of the initial universal quantifier $(\forall x)$ (with an index $(\forall x)_1$) consists of two segments of the formula in question. The former is

$$_1((\exists y)(\forall z))_1$$

and the latter is

$$_1(S[x,y,u,z])_1$$

where in both segments the subscripts merely indicate that the parentheses in question belong to $(\forall x) = (\forall x)_1$. In contrast, the scope of $(\forall z) = (\forall z)_2$ is a continuous one, consisting of

$$_2((\exists u)S[x,y,z,u])_2$$

But by thus lifting the quantifier $(\exists u)$ from the scope of $(\forall x)$ we accomplish precisely the same as is done in my notation by writing it as $(\exists u/\forall x)$.

This shows how fundamental Frege's fallacy was. It could be corrected even without introducing any new symbols into our logic, by merely changing the use of the innocent-looking device of parentheses. (The subscripts are not any more meaningful in themselves than connecting lines.) At the same time (3.23) shows that such a procedure would be so clumsy as to be impracticable.

No matter which notation is used, the step from ordinary first-order logic to IF first-order logic is little more than the recognition of the fact that the two kinds of scope do not have to go together. For instance, in our slash notation the formal scope (indicated by parentheses) can be taken to be the binding scope, while the slash notation involves a recognition of the fact that priority scope does not automatically follow the binding scope.

(iii) A special case of the phenomenon of informational independence has been studied by logicians ever since Henkin (1961) under the heading "branching quantifiers" (cf. e.g., Barwise 1979, Enderton 1970, Walkoe 1970). The connection between the two can be seen from an example. As was noted previously, our familiar Henkin quantifier sentence

$$(\forall x)(\forall z)(\exists y/\forall z)(\exists u/\forall x)S[x,y,z,u] \tag{3.12}$$

can be written, instead of the slash notation, in a nonlinear notation as follows:

$$\left.\begin{array}{l}(\forall x)(\exists y)\\(\forall z)(\exists u)\end{array}\right\} S[x,y,z,u] \tag{3.9}$$

Here the mutual dependencies and independencies of different quantifiers are indicated by a partial ordering of those quantifiers. This idea can easily be generalized. However, it does not yield a really satisfactory analysis of the situation. For one thing, the branching notation has not been used in connection with propositional connectives, which can also exhibit informational independence, until very recently and for the first time apparently in Sandu and Väänänen (1992).

Also, the partial ordering of quantifiers corresponds in a game-theoretical approach to a partial ordering of their information sets with respect to class-inclusion. But in general there is no reason whatever to think that the information sets of different moves in a game are even partially ordered. Hence branching or partially ordered quantifiers are not representative of the general situation. Admittedly, it turns out that in the special case of quantifiers all their possible patterns of dependence and independence do reduce to a suitable partial ordering. However, this is not true in general.

Furthermore, one has to handle negation very cagily in the context of informationally independent quantifiers, as will be shown in Chapter 7. It is not even clear that all the earlier treatments of branching quantifiers handle negation in a consistent manner.

Most importantly, branching is a matter of the interplay of quantifiers, and not of the interpretation of quantifiers taken one by one. For this reason, generalizations of the notion of quantifier taken in isolation from other quantifiers do not help essentially in understanding branching and other informationally indpendent quantifiers. For instance, in spite of the other kinds of success that the theory of generalized quantifiers has enjoyed, it has not helped essentially to understand branching quantifiers. Admittedly, there exist treatments of informationally independent quantifiers in the tradition of generalized quantifiers (see, for example, Westerstahl 1989 and Keenan and Westerstahl 1996). They do not bring out the generality of the phenomenon of independence, however, or even reach a fully satisfactory treatment of the semantics of informationally independent quantifiers.

However, some of the general issues that are prompted by the phenomenon of informational independence have been discussed by philosophers and logicians in the earlier literature in the special case of branching quantifier structures. Among other things, several of the most important perspectives of IF first-order languages are generalizations of the properties of branching quantifiers. I will return to them later in this chapter.

(iv) The best measure of the naturalness of IF first-order logic is the simplicity of the (game-theoretical) semantics that governs it. How are the semantical game rules for IF first-order languages related to those of ordinary first-order languages? The relationship is an interesting one – and a simple relationship at that. It is one of identity. The rules for making moves in a semantical game in IF first-order logic are precisely the same as those used in ordinary first-order logic, except that imperfect information is allowed. The extensive form of a game connected with an IF sentence is the same as that of the corresponding ordinary first-order formula, except for the fact that the information sets of some of the moves do not contain all the earlier moves. (The information set of a move includes all and only those earlier moves which are known to the player making the move.)

Moreover, it is important to realize that this failure of perfect information has nothing subjective about it. It is not a matter of a limitation of human knowledge or human ingenuity. It is a feature of the combinatorial situations that can come up in a semantical game. The truth or falsity of a sentence S of an IF language is an objective fact about the world which S can be used to speak of. This objectivity of truth in IF first-order logic will be further illustrated in Chapter 7.

(v) It is to be noted that the irreducibly independence-friendly formulas are typically stronger than the corresponding formulas of traditional first-order logic. The reason is clear from a game-theoretical perspective. From this viewpoint, a sentence asserts the existence of a winning strategy for the initial verifier in the corresponding game. The presence of informational independence means the use of fewer arguments in some of the strategy functions; in other words, the use of a strategy based on less information. Hence the existence of such a strategy is a stronger claim than the existence of the analogous strategy based on more information. It thus turns out that IF first-order languages are not only stronger than ordinary ones, but extremely strong.

The greater expressive strength of IF first-order logic as compared to its older brother has a concrete meaning. It does not mean only, or in the first place, greater deductive resources. What it means is that there are classes (kinds) of structures (even finite ones) that simply cannot be described by means of the resources of ordinary first-order logic. Particular examples will be discussed in Chapter 7. This kind of advantage of IF first-order logic is naturally of great potential interest both theoretically and in practice – that is, in view of the role of logic in mathematical theorizing.

An example may help the reader to appreciate this point. Consider the following formula

$$(\forall x)(\forall z)(\exists y/\forall z)(\exists u/\forall x)(((x = z) \leftrightarrow (y = u))$$
$$\& \ H(x, y) \ \& \ H(z, u) \ \& \ \sim H(x, u) \ \& \ \sim H(z, y)) \qquad (3.24)$$

It is easily seen to be equivalent to

$$(\exists f)(\forall x)(\forall z)(((x \neq z) \supset (f(x) \neq f(z))) \ \& \ H(x, f(x))$$
$$\& \ H(z, f(z)) \ \& \ \sim H(x, f(z)) \ \& \ \sim H(z, f(x))) \qquad (3.25)$$

In other words, what (3.24) says is that one can correlate with each x an individual $f(x)$ such that the correlates of two different individuals are different and differ from each other in the way indicated by (3.25). Thus, if $H(x, y)$ is read "y is a hobby of x's", then (3.24) says something like "everybody has a unique hobby."

It can be shown that (3.24)–(3.25) do not reduce to any expressions of an ordinary first-order language. What (3.24) does is therefore to capture as its models a class of structures with a relatively simple characteristic common feature. Yet this class of structures cannot be captured by ordinary first-order logic. Thus the basis of the greater strength of IF first-order logic is its greater representational capacity, and not its ability to offer to a logician new deductive gimmicks.

(vi) In spite of their strength, IF first-order languages have a number of welcome properties. Among other results, IF first-order logic has the following properties:

(A) It is compact. This fact is easily proved. Let σ be an infinite set of formulas of IF first-order logic. Let σ^* be the set of second-order translations of all the members of σ, with different initial existential function quantifiers using different function variables. Let σ^{**} be the result of omitting all these initial function quantifiers from all the members of σ^*. This σ^{**} is a set of ordinary first-order formulas. Clearly, if one of σ, σ^* and σ^{**} is satisfiable, then all these are. Because ordinary first-order logic is compact, σ^{**} is consistent if and only if all its finite subsets are consistent. But clearly such a subset τ^{**} is consistent if and only if its parent subset τ of σ is consistent. And τ is consistent by hypothesis.

(B) IF first-order logic has the (downwards) Löwenheim–Skolem property. In order to prove this let σ be a finite set of formulas of an IF first-order language, and let σ^* and σ^{**} be formed as before. Clearly σ, σ^* and σ^{**} are satisfiable in the same domains. But in virtue of the downwards Löwenheim–Skolem theorem for ordinary first-order logic, σ^{**} is satisfiable in a countable domain if satisfiable at all.

(C) The separation theorem holds in a strengthened form. To see this, assume that σ and τ are sets of formulas of an IF first-order language satisfying the following conditions:

(i) σ and τ are both consistent.

(ii) $\sigma \cup \tau$ is inconsistent.

Then there is a formula of the corresponding ordinary first-order language (i.e., a slash-free formula) S such that

(iii) $\tau \vdash S \quad \sigma \vdash \sim S$

(iv) All the nonlogical constants of S occur in the members of both σ and τ.

The separation theorem can be proved by transforming σ and τ into σ^* and σ^{**} as well as into τ^* and τ^{**} as before. It is easily seen that the separation theorem of ordinary first-order logic applies to σ^{**} and τ^{**}, yielding an ordinary first-order formula S_0 as the separation formula. This S_0 is easily seen to serve also as the separation formula claimed to exist in the theorem.

(D) IF first-order languages admit of the Skolem normal form in a stronger sense than ordinary first-order languages. In them, the Skolem normal form S_2 of a given formula S_1, which is a formula of the form

$$(\forall x_1)(\forall x_2)\ldots(\exists y_1)(\exists y_2)\ldots S[x_1, x_2, \ldots, y_1, y_1, \ldots] \qquad (3.26)$$

is satisfiable if and only if S_1 is satisfiable. But S_1 and S_2 need not be logically equivalent nor have the same vocabulary. But in IF first-order logic, each formula can easily be seen to have an equivalent formula which is like (3.26) except that every $(\exists y_k)$ is replaced by

$$(\exists y / \forall x_{k1}, \forall x_{k2}, \ldots) \qquad (3.27)$$

where

$$\{x_{k1}, x_{k2}, \ldots\} \subseteq \{x_1, x_2, \ldots\} \qquad (3.28)$$

The given formula S_1 can be transformed into S_2 by first transforming it into its prenex form, that is, into a form where all quantifiers are at the beginning of the formula. Then one can extend the scope of a universal quantifier $(\forall x)$ over a number of existential quantifiers $(\exists y_1), (\exists y_2), \ldots$ if one replaces them by $(\exists y_1 / \forall x), (\exists y_2 / \forall x), \ldots$. The same can be done to several universal quantifiers at the same time. This yields the desired form.

In results like (d) we can also include within their scope the initially tacit "Skolem functions" associated with disjunctions. For instance, the Skolem normal form of

$$(\forall x)(\exists y)((\forall z)S_1[x, y, z](\vee / \forall x)S_2[x, y]) \qquad (3.29)$$

can be

$$(\forall x)(\forall z)(\exists y/\forall z)(\exists u/\forall x, \forall z)((S_1[x, y, z]\&(u=0))$$

$$\vee (S_2[x, y] \& (u \neq 0))) \tag{3.30}$$

(E) Beth's theorem on definability holds in IF first-order logic in a sharper form. This theorem says that an "implicit" definability of a constant, say a one-place predicate P, on the basis of a consistent theory $T[P]$, entails its explicit definability on the basis of the same theory. In other words, assume that

$$(T[P]\&T[P'])\vdash(\forall x)(P(x)\leftrightarrow P'(x)) \tag{3.31}$$

where P' is a new one-place predicate constant. Then for a suitable definiens (complex formula) $D[x]$ with x as its only individual variable but without P (or P') we have

$$T[P]\vdash(\forall x)(P(x)\leftrightarrow D[x]) \tag{3.32}$$

This result can be proved for IF first-order logic on the basis of the separation theorem in precisely the same way as the corresponding result for ordinary first-order logic. From the proof (and the separation theorem) it can be seen that the definiens $D[x]$ can always be chosen to be a formula of ordinary first-order logic.

(F) Just as every ordinary first-order sentence was seen in Chapter 2 to have a second-order translation, in the same way each sentence of an IF first-order language can be seen to have a similar second-order translation. The translation is obtained in the same way as in ordinary first-order languages (see Chapter 2). The only difference is that the arguments $x_1, x_2, \ldots$ are dropped from the function correlated with $(\exists y/\forall x_1, \forall x_2, \ldots)$ or with $(\vee /\forall x_1, \forall x_2, \ldots)$.

(G) As in ordinary first-order logic, the second-order translations of IF first-order sentences are of the Σ_1^1 form. Unlike ordinary first-order logic, however, IF first-order logic also allows for the converse translation. In other words, each Σ_1^1 sentence has a translation (logical equivalent) in IF first-order language. Even though this result is known from the literature, it is still useful to see how it can be established.[1]

A Σ_1^1 sentence has the form of a sequence of second-order existential quantifiers followed by a first-order formula. First, we replace all predicate variables by function variables. This can be done one by one.

For instance, suppose we have a formula of the form

$$(\exists X)S[X] \tag{3.33}$$

where X is a one-place predicate variable. Then we can replace (3.33) by the Σ_1^1 formula

$$(\exists f)S'[f] \tag{3.34}$$

where $S'[f]$ is obtained from $S[X]$ by replacing every atomic formula of the form $X(y)$ or $X(b)$ by

$$f(y) = 0 \tag{3.35}$$

or

$$f(b) = 0 \tag{3.36}$$

respectively.

After eliminating all predicate variables in this way, all the formulas we need to deal with are of the form

$$(\exists f_1)(\exists f_2)\ldots(\exists f_k)(\forall x_1)(\forall x_2)\ldots(\forall x_l)S[f_1, f_2, \ldots, f_k, x_1, x_2, \ldots, x_l] \tag{3.37}$$

where $S[\ ]$ is a quantifier-free ordinary first-order formula.

Here (3.37) can be expressed in an equivalent form in the IF first-order notation if the following two conditions are satisfied:

(a) There is no nesting of functions f_i.

(b) Each f_i occurs in $S[\ ]$ with only one sequence of arguments, say, with the variables $x_{i1}, x_{i2}, \ldots$, missing from its set of arguments.

For then we can replace (3.37) simply by

$$(\forall x_1)(\forall x_2)\ldots(\forall x_l)\ldots(\exists y_i/\forall x_{i1}, x_{i2}, \ldots)S^* \tag{3.38}$$

where S^* results from replacing each f_i followed by its arguments by y_i.

What has to be shown is therefore how an arbitrary (3.37) can be brought to a form where (a)–(b) are satisfied.

I will illustrate this transition by means of two examples. First, consider a formula of the form

$$(\exists f)(\forall x_1)(\forall x_2)S[f(x_1), f(x_2)] \tag{3.39}$$

Here (a) is satisfied. In order to bring it to a form where (b) is also satisfied, we can rewrite it as

$$(\exists f)(\exists g)(\forall x_1)(\forall x_2)(((x_1 = x_2) \supset (f(x_1) = g(x_2))) \& S[f(x_1), g(x_2)]) \tag{3.40}$$

This is easily seen to be equivalent to

$$(\forall x_1)(\forall x_2)(\exists y_1/\forall x_2)(\exists y_2/\forall x_1)(((x_1 = x_2) \supset (y_1 = y_2)) \ \& \ S[y_1, y_2])$$
(3.41)

For the second example, consider a formula of the form

$$(\exists f_1)(\exists f_2)(\forall x)S[f_2(f_1(x))]$$
(3.42)

Here condition (a) is not satisfied. In order to meet this condition, we can consider the formula

$$(\exists f_1)(\exists f_2)(\forall x_1)(\forall x_2)(\forall x_3)(((x_2 = f_1(x_1)) \ \& \ (x_3 = f_2(x_2))) \supset S[x_3])$$
(3.43)

which satisfies both (a) and (b). Indeed, (3.43) is logically equivalent to

$$(\forall x_1)(\forall x_2)(\forall x_3)(\forall x_4)(\exists y_1/\forall x_2\forall x_3\forall x_4)(\exists y_2/\forall x_1\forall x_3\forall x_4)$$
$$(\exists y_3/\forall x_1\forall x_2\forall x_4)(\exists y_4/\forall x_1\forall x_2\forall x_3)$$
$$(((x_1 = x_3)\leftrightarrow(y_1 = y_3)) \ \& \ ((x_2 = x_4)\leftrightarrow(y_2 = y_4)) \ \&$$
$$(((y_2 = x_2) \ \& \ (y_2 = x_3)) \supset S[y_2]))$$
(3.44)

In neither example are the other singular terms occurring in S affected. We can therefore eliminate one by one all the violations of the requirements (a) and (b) mentioned earlier in this section. This process terminates after a finite number of steps, yielding the desired translation.

(vi) At this point, it is in order to look back at the precise way the information sets of different moves are determined in semantical games with IF sentences. The small extra specification that is needed is that moves connected with existential quantifiers are always independent of earlier moves with existential quantifiers. This assumption was tacitly made in the second-order translation of IF first-order sentences explained earlier in this chapter. The reason for this provision is that otherwise "forbidden" dependencies of existential quantifiers on universal quantifiers could be created through the mediation of intervening existential quantifiers.

This point can be illustrated as follows: If the provision is heeded, the second-order translation of

$$(\forall x)(\forall z)(\exists y/\forall z)(\exists u/\forall x)S[x, y, z, u]$$
(3.45)

will be

$$(\exists f)(\exists g)(\forall x)(\forall z)S[x, f(x), z, g(z)] \qquad (3.46)$$

If it is not heeded, the translation will be

$$(\exists f)(\exists h)(\forall x)(\forall z)S[x, f(x), z, h(z, f(x))] \qquad (3.47)$$

The two second-order translations (3.46) and (3.47) are not equivalent. Obviously (3.46) is at least as strong as (3.47). Without the provision, there is no way of capturing the stronger statement (3.46) in IF first-order logic. Hence the extra provision has to be adopted.

I will not try to prove this impossibility here – that is to say, the impossibility of expressing (3.46) in an IF notation without the convention of omitting the initial verifier's moves from information sets. What can be easily shown is that we get a different interpretation with and without the convention. For the purpose, it suffices to provide an example where the second-order translations (3.46) and (3.47) are not equivalent. Such an example is

$$(\forall x)(\forall z)(\exists y/\forall z)(\exists u/\forall x)((y \neq x)\&(u \neq z) \,\&\, ((x = z)\leftrightarrow(y = u)) \quad (3.48)$$

With the convention, the nonempty models of (3.48) are all infinite ones. Without the convention, (3.48) is satisfied in the following two-element model **M**:

$$do(\mathbf{M}) = \{a, b\} \qquad (3.49)$$

$$
\begin{aligned}
&f(a) = b \quad f(b) = a \\
&g(a, b) = b \quad g(b, a) = a \\
&g(a, a) = b \quad g(b, b) = a
\end{aligned}
$$

It is easy to see that f and g satisfy the formula

$$(\forall x)(\forall z)((x = z)\leftrightarrow(f(x) = g(z, f(x)))) \qquad (3.50)$$

which is a subformula of the translation of (3.48) to a second-order form without the convention. This translation is of course

$$(\exists f)(\exists g)(\forall x)(\forall z)((x = z)\leftrightarrow(f(x) = g(z, f(x)))) \qquad (3.51)$$

All this illustrates the indispensability of the convention for a reasonable semantics of IF first-order logic.

(viii) It is important to understand precisely what the relationship is between IF first-order logic and its traditional order brother. The following are among the most relevant aspects of this relationship:

(a) Ordinary first-order logic is a special case of IF first-order logic.

(b) No ideas are involved in IF first-order logic that were not needed for the understanding of ordinary first-order logic.

(c) Technically speaking IF first-order logic is a conservative extension of ordinary first-order logic.

IF first-order languages nevertheless differ strikingly from the ordinary first-order logic in several ways. Three of them are so important as to serve naturally as lead-ins into the next three chapters. These three forms of unfamiliar logical behavior are the following:

(a) IF first-order logic does not admit of a complete axiomatization.

(b) IF first-order logic does not admit of a Tarski-type truth definition.

(c) The law of excluded middle does not hold in IF first-order logic.

Combined with the status of IF first-order logic as the natural basic core area of logic, these observations will be seen to put the entire foundations of logic in a new light, as I will show in the subsequent chapters of this book. Among the minor new bridges one has to cross here is the precise definition of notions like logical consequence and logical equivalence. The novelty is due to the failure of the law of excluded middle, which forces us to associate with each sentence S not only the class $M(S)$ of all models in which it is true, but also the class $\bar{M}(S)$ of all models in which it is false, for the latter is no longer a mere complement of the former. This prompts the question: Are two sentences logically equivalent if and only if they are true in the same models (on the same interpretation), or are they to be so called if and only if they are true and false in the same models? In this work, I will consistently opt for the former alternative, and for its analogies with other concepts of the metatheory of logic.

Indeed, an Argus-eyed reader may already have noticed that I tacitly relied on such an understanding of the basic metalogical notions like logical equivalence in the preceding arguments that established some of the basic logical properties of IF first-order languages. Moreover, and most importantly, an IF first-order sentence and its second-order translation are true in the same models, but they are not false in the same models. This follows from the fact that *tertium non datur* holds in ordinary second-order logic, but not in IF first-order logic.

By means of IF first-order logic we can overcome most of the limitations of ordinary first-order logic which were pointed out in earlier chapters. IF first-order logic is thus the key tool in my approach to the foundations of logic and mathematics. Earlier, I argued for the status of IF first-order logic as the core area of all logic. This special status of the notion of informational independence can be illustrated further by pointing out the ubiquity of the phenomenon of informational independence in the semantics of natural languages. This ubiquity is not obvious prima facie, although the success of GTS makes it predictable. The reasons for its inconspicuousness are diagnosed in Hintikka (1990).

(ix) Some philosophers have seen in the absence of a complete proof procedure the basis of a serious objection to IF first-order logic. Such a way of thinking is based on a number of serious mistakes. Apparently the skeptics think that they cannot understand a language unless they understand its logic; that they do not understand its logic unless they understand what its logical truths are; and that they do not understand the logical truths of a part of logic unless they have a complete axiomatization of such truths (valid formulas). The last two steps of this line of thought are both fallacious. First, what is needed to understand a language is to understand the notion of truth (truth *simpliciter*, not logical truth) as applied to it. The reason is that what a sentence says is in effect that the world is such that it is true in the world. This point is completely obvious, and it has been emphasized by a variety of prominent philosophers all the way from Frege to Davidson. I do not see much point in arguing for it elaborately here.

Since IF first-order languages admit truth definitions expressible in the language itself, as will be explained in a later chapter, they cannot possibly be faulted on this point.

The subtler mistake seems to be that in order to understand logic one has to understand its logical truths. This is at best a half-truth. One possible mistake here is to think that logical truths are merely a subclass of truths. Hence, according to this line of thought, a satisfactory account of truth for a language ought to yield as a special case an account of logical truth for this language. But logical truth simply is not a species of (plain) truth, notwithstanding the views of Frege and Russell. Logical truths are not truths about this world of ours. They are truths about all possible worlds, truths on any interpretation of nonlogical constants. This notion is an animal of an entirely different color from ordinary truth in some one world. In this sense, to call valid sentences "logically *true*" is a misnomer. Several philosophers who swear by the ordinary notion of truth are nonetheless more than a little skeptical of the ultimate viability of the very notion of logical truth.

Moreover, it simply is not true that the only way of coming to master the notion of logical truth is to provide a recursive enumeration of logically true sentences. Such an enumeration is neither necessary nor sufficient for the purpose. On the one hand, even if we can give such an enumeration, it is not automatically clear what it is that is being enumerated. On the other hand, there are other ways than recursive enumeration of coming to understand the idea of logical truth. Indeed, if you understand the ordinary notion of truth in a model and also understand what the totality of all models is, you can characterize valid (logically true) sentences that are true (in the ordinary sense) in all models. This way of coming to understand the notion of logical truth does not presuppose the axiomatizability of all logical truths nor indeed any logical proof method at all.

It is fairly clear that many of the philosophers who object to IF first-order logic on the grounds of its incompleteness are assuming some version of the syndrome which I have called belief in the universality of language. Among its manifestations is the assumption that all semantics is ineffable and the thesis that properly speaking we can speak of only *this* world of ours. If the latter thesis is true, logical truths must be a subclass of ordinary truths, as Frege and Russell thought, for there is nothing else for them to be true about. Whatever one thinks of such theses, it is inappropriate to base one's arguments on them here. The reason is that the question of the definability or undefinability of truth is one of the most important test cases as to whether the universality assumption is correct or not. Hence to assume universality in the present instance is in effect to argue in a vicious (or at best uninformative) circle.

Some philosophers, for instance Quine, have expressed the fear that a logic which is not axiomatizable is too unwieldy to be viable. IF first-order logic offers a counterexample to such paranoia. How manageable IF first-order logic is conceptually, is attested to by the fact that most of the "nice" metatheorems that hold in ordinary first-order logic also hold in IF first-order logic, as we have seen.

(x) The apprehensions prompted by the nonaxiomatizability of the logical truths of IF first-order logic can perhaps be partly allayed by considering the question of the axiomatizability of the inconsistent (unsatisfiable) formulas of IF first-order logic. A moment's thought shows that the class of inconsistent formulas of our new logic is indeed axiomatizable. This can be shown – and an axiomatization can be found – along the usual way. Given a finite set σ of formulas, we can formulate a set of rules for adding new formulas to it along the same lines as, say, in the step-by-step rules for constructing a model set alias "Hintikka set" in the sense of Smullyan (1968). Intuitively, we can think of following these rules as an attempt to describe more and more fully a model in which all the members of σ are true. The rules for doing so are, formally speaking, inverses of certain disproof rules. It is not difficult to show (though I will not do so here) that these rules can be chosen so as to be complete. That is to say, they can be chosen in such a way that they bring out the inconsistency of any inconsistent σ that can be uncovered by a finite number of applications of the construction rules: all the different ways of trying to construct a model for the members of σ will run into a dead end, just as in ordinary first-order logic.

The only reason why this complete disproof procedure does not yield a proof procedure in IF first-order logic is the behavior of negation. For a complete proof procedure, say, a procedure for proving a conclusion C from a set σ of premises, we need rules for trying to construct a model in which the members of σ are true but C not true. In general, alas, there is no contradictory negation of C in the relevant IF first-order language (i.e., no formula which would be true just in case C is not) for which we could try to construct a model jointly with σ. It is for this reason that a complete disproof procedure does not translate into a complete proof procedure in the teeth of a failure of a *tertium non datur*.

Most of the details of a disproof procedure of the kind just envisaged are not interesting. There nevertheless is one point worth noting in it. In the usual proof and disproof procedures, the lion's share of the work is done by instantiation rules or by rules equivalent to them, especially by the rule of existential instantiation or its dual, universal generalization. In the usual simple form of the rule of existential instantiation, a new individual constant is brought in to replace the variable bound to an existential quantifier.

In IF logic, such a simple rule of existential instantiation is no longer adequate. Such a rule cannot discriminate different kinds of dependencies of an existential quantifier on universal quantifiers with a wider scope. Instead of the former rule which simply takes us from an existentially quantified sentence

$$(\exists x)S[x] \tag{3.52}$$

to its substitution-instance

$$S[b] \tag{3.53}$$

where b is a new individual constant, we must have a rule which takes us from a formula

$$S_1 = S_1[(\exists y/\forall x_1, \forall x_2, \ldots)S_2[y]] \tag{3.54}$$

(in the negation normal form) which contains an existentially quantified subformula to a new sentence of the form

$$S_1[S_2[f(z_1, z_2, \ldots)]] \tag{3.55}$$

where $(\forall z_1), (\forall z_2), \ldots$ are all the universal quantifiers other than $(\forall x_1), (\forall x_2), \ldots$ within the scope of which $(\exists y/\forall x_1, \forall x_2, \ldots)$ occurs in S_2 and where f is a new function constant. This rule might be called the rule of functional instantiation. We also need a stipulation to the effect that a rule of universal instantiation must not be applied to a formula (in negation normal form) to which the rule of functional instantiation can be applied. Otherwise we cannot reach a sound and complete system of rules of disproof for IF first-order logic along anything like the usual lines.

This result might seem to be rather quaint, but on a closer scrutiny it turns out to throw light on general issues concerning the nature of quantifiers. If the idea of quantifiers as higher-order predicates is right, then a first-order existential quantifier prefixed to an open formula says merely that the (usually complex) predicate defined by that open formula is not empty. The resulting logic so understood must be exhausted by the usual rule of existential instantiation which introduces one new individual constant, for the nonemptyness of a predicate can be expressed by using such a constant. However, we just saw that such a simple instantiation rule is not sufficient or always admissible. Why not? The reason obviously is that Hilbert was right and Frege wrong – that is, that quantifiers are in reality tacit choice functions that depend on some outside universal quantifiers but not on others. The insufficiency of the usual rule of existential instantiation in IF first-order logic thus constitutes telling evidence against the idea that quantifiers are merely higher-order predicates.

The naturalness of the rule of functional instantiation can be illustrated in a variety of ways. For instance, a perceptive reader may have noticed that it is very closely related to the way most meta-theoretical results were proved, as in Section (a). Furthermore, the old-fashioned rule of existential instantiation can be thought of as a limiting case of functional instantiation and the instantiating individual constant as a function constant with zero arguments. Thus the sufficiency of the rule of existential instantiation in ordinary first-order logic can be thought of as yet another piece of undeserved luck which has befallen this particular logic, but which is not representative of the real situation in logic in general.

Conversely, the new rule of functional instantiation is a most natural generalization of the old rule of existential instantiation. Other changes in the familiar rules of logical inference may likewise be needed in IF first-order logic. For instance, we may choose to stipulate that the inclusion of the syntactical scope of $(Q_2 x)$ within the scope of $(Q_1 x)$ automatically means that the former depends on the latter. If so, the usual rules for extending and contracting the scope of the quantifier over another quantifier may have to be reformulated.

Another dimension of the need for a rule of functional instantiation is that it illustrates the failure of compositionality in IF first-order logic. In the most natural rules of disproof for this logic, what is substituted for an existentially bound variable depends for its context on the entire initial formula to be tested for possible inconsistencies.

Yet another implication of the need for a functional instantiation rule pertains to the relation of the semantical games of verification and falsification to the games of formal proof and disproof. At first sight, it looks as if one could read from the rules of verification games rules for constructing a world in which there exists a winning strategy for the initial verifier. In a certain sense this holds for disproof procedures, but not in as direct a sense as might first seem to be the case. For the existence of a winning strategy is not a matter of the possibility of making moves in a certain way independently of other moves. The existence of a strategy involves dependencies of moves on or of each other. Such (in)dependencies can be captured only by means of functions.

Thus in a very real sense IF logic does not only force us to rethink the semantics of our basic logic; it forces us also to reconsider our rules of logical inference. In terms of my anachronistic indictment of Frege, his mistake is not limited to formulating his formation rules too restrictively. If he had removed the restrictions, he would also have had to strengthen his rules of inference.

Note

[1] This seems to have been established first by Walkoe (1970); but see also Enderton (1970) and Barwise (1979).

4

The Joys of Independence:
Some Uses of IF Logic

So far, I have been talking about what IF logic is like. The next question is what such a logic can do. This question admits of a variety of partial answers already at the present stage of the exploration that is being carried out here.

Perhaps the most general key to the usefulness of a recognition of informational independence is the ubiquity of such independence. Many philosophers still seem to think of branching quantifiers – and informational independence in general – as a marginal phenomenon, a logician's curiosity without deeper theoretical interest. In view of such views, it is important to realize that informational independence is in reality a widespread and important feature of the semantics of natural languages. It can be shown to play a crucial role in epistemic logic, in the theory of questions and answers, in the *de dicto* versus *de re* distinction, and so forth.

The prevalence of informational independence does not come as a surprise to anyone who thinks game-theoretically about the semantics of natural languages. The leading idea of such a game-theoretical approach is to associate a game rule to each different structural feature and to each lexical item of a natural language. But as soon as that has been done, you have – or, rather, language users have – the option of considering some moves as being made on the basis of nonmaximal information sets, no matter what the structural feature or lexical item in question is. Hence the prevalence of independence is virtually a corollary of the basic idea of GTS, assuming of course its success in elucidating the different aspects of the semantics of natural languages.

The ubiquity of the phenomenon of informational independence in natural language nevertheless has been hidden by the fact that informational independence is not indicated in natural languages by any uniform syntactical device. The reasons for this secretiveness of natural language grammar turn out in fact to be highly interesting from the viewpoint of linguistic theorizing (see Hintikka 1990).

A terminological comment may be in order here. At least one observer has in discussion called IF first-order logic "deviant". The epithet is undeserved. IF first-order logic is a conservative extension of ordinary first-order logic. It contains classical first-order logic as a part, and hence cannot be called nonclassical except on the pain of deviant terminology. More than that: even though the extension beyond ordinary first-order logic which IF first-order logic represents exhibits nonclassical behavior, this extension is forced on us by perfectly classical principles. What happened was that I assumed some perfectly classical laws for the nonordinary part of IF first-order logic, such as DeMorgan's laws, the law of double negation, and so forth. But then it turned out that those assumptions entailed a violation of another allegedly classical law, namely, the law of excluded middle. Hence attributes like "deviant" and "nonclassical" when applied to IF first-order logic are apt to direct a hearer's attention to a diametrically wrong direction.

One can also point out that, in principle, interpreted IF first-order sentences are not connected with experience in a more distant manner than ordinary first-order sentences. I have pointed out on earlier occasions, for instance in Hintikka (1988a), that even sentences with a complex quantifier prefix can be thought of as having been learned directly. In particular, what a controlled experiment shows in a simple two-variable case is how the observed variable depends on the control variable. The outcome of such an experiment is therefore a sentence of the form

$$(\forall x)(\exists y)S[x, y] \tag{4.1}$$

normally with some restriction $x_1 < x < x_2$ on x. More complex experimental setups can yield more complex propositions. A moment's thought shows that they can likewise yield an IF first-order sentence, for example,

$$(\forall x)(\forall z)(\exists y/\forall z)(\exists u/\forall x)S[x, y, z, u] \tag{4.2}$$

Here x and z are the control variables, y and u the observed ones. What the experiment will have to show is, among other things, that y does not depend on z or u on x. Establishing such an independence experimentally may be difficult in practice, but it is not impossible in principle.

One can say more here, however. One surprise is not enough. It was seen in the preceding chapter that things are not in logic as we used to think. The true basic logic is not ordinary first-order logic, but independence-friendly first-order logic. But things are not as we used to think in mathematics, either. The usual picture of mathematical practice is of a mathematician carrying out proofs in terms of first-order logic. Whatever assumptions one has to make beyond first-order logic are set-theoretical, and the entire enterprise can in principle be carried out within an axiomatic set theory.

This is a misleading picture of mathematics. One reason why it is misleading is that mathematics is full of conceptualizations which in the last analysis must be expressed by means of IF first-order logic rather than of ordinary first-order logics, even though this ingredient is seldom acknowledged in so many words. Perhaps the most familiar example is offered by the notion of uniform differentiability. The function $f(x)$ is differentiable at each point of an interval $x_1 < x < x_2$ if and only if

$$(\forall x)(\exists y)(\forall \varepsilon)(\exists \delta)(\forall z)(((x_1 < x < x_2) \,\&\, (|z| < |\delta|))$$

$$\supset (|((f(x+z) - f(x))/z) - y| < |\varepsilon|)) \tag{4.3}$$

The function is uniformly differentiable in the same interval if and only if the same condition is satisfied with $(\exists \delta)$ replaced by $(\exists \delta / \forall x)$. In this case of uniform differentiability, a confusion between dependent and independent quantifiers has actually manifested itself in the history of mathematics (see, for instance, Grabiner 1981, p. 133).

In general, the mathematical notion of uniformity is closely related to the idea of informational independence.

Such examples are easily multiplied. In many cases, the use of IF quantifiers is hidden by the use of function symbols. Consider a second-order formula where two functions are asserted to exist. Suppose the argument sets of those functions are not linearly ordered by class-inclusion. Then the resulting expression will have the logical form

$$(\exists f_1)(\exists f_2) \cdots (\forall x_1)(\forall x_2) \cdots S[f_i(x_{i1}, x_{i2}, \ldots)] \tag{4.4}$$

where the sets $\{x_{i1}, x_{i2}, \ldots\}$ are not linearly ordered by class inclusion. Then the translation of (4.4) to the first-order level will normally involve irreducibly IF quantifiers.

Even a sentence asserting the existence of a single function, whose argument sets at its different occurrences are not linearly ordered by class inclusion, can be a case in point. An example might be of the form

$$(\exists f)(\forall x)(\forall z)S[x, z, f(x), f(z)] \tag{4.5}$$

Here (4.5) is easily seen to be equivalent to

$$(\forall x)(\forall z)(\exists y/\forall z)(\exists u/\forall x)(S[x, z, y, u] \mathbin{\&} ((x = z) \supset (y = u))) \tag{4.6}$$

Such examples perhaps illustrate why informational independence is not an unknown phenomenon in working mathematics. One cannot perhaps hope to find examples of branching first-order quantifier structures explicitly mentioned in mathematical treatises, but one should not be surprised to find mathematical results which assert the existence of two functions of one variable whose values for respectively different arguments can be related to each other in a specifiable way, as in a sentence like

$$(\exists f)(\exists g)(\forall x)(\forall z)S[x, f(x), z, g(z)] \tag{4.7}$$

And, as has been seen, to assert (4.7) is equivalent to asserting the independence-friendly first-order sentence

$$(\forall x)(\forall z)(\exists y/\forall z)(\exists u/\forall x)S[x, y, z, u] \tag{4.2}$$

The connection between the notion of informational independence and the relation between the argument sets of different functions can be put to use for clarificatory purposes that transcend the scope of logic and mathematics. Consider, as an example, the question as to when two physical systems Σ_1 and Σ_2 are independent of each other. Let us assume, for the sake of analysis, that these systems and their behavior are specified in first-order terms. Then one may be tempted to say that Σ_1 and Σ_2 are independent if and only if the quantifiers characterizing one range over a set of individuals different from, and exclusive of, the range of values of quantifiers characterizing the other. We can now see that this requirement is not always sufficient. For the relative order of the members of the two sets of characterizing quantifiers may still create dependencies between the two sets, and hence between the two systems. Hence, for a satisfactory characterization of the mutual independence of the two systems, it must also be required that the quantifiers characterizing one are informationally independent of those characterizing the other.

An especially interesting and tacit use of informationally independent quantifiers is found in the mathematics of quantum theory. Let us assume that we are there considering a systems whose state depends among others on two variables x, z. Let y and u be variables for the observed values of x and z, respectively. Let $S[x, y, z, u]$ be a specification of how x, y, z, u are related to each other. In practice, $S[x, y, z, u]$ includes a specification of the state of the system in question, including the values of x and z. It will also include a specification of the projection operators which lead from the state vector to the possible observed values y and u.

The crucial question is whether x and z are simultaneously measurable. This question means asking whether y can depend only on x and u only on z. Otherwise, the observed values cannot help to specify the actual ones. But this question amounts to asking whether the following sentence is true:

$$(\exists f)(\exists g)(\forall x)(\forall z)S[x, f(x), z, g(z)] \qquad (4.7)$$

If x and z are conjugate variables, the well-known answer is that (4.7) is not true. Of course, this does not preclude that one can measure either x or z. However, if one then tries to measure the other one, too, the result depends also on the first variable. In other words, the following sentences can be true even though (4.7) is not:

$$(\exists f)(\exists g)(\forall x)(\forall z)S[x, f(x), z, g(z, x)] \qquad (4.8)$$

$$(\exists f)(\exists g)(\forall x)(\forall z)S[x, f(x, z), z, g(z)] \qquad (4.9)$$

But this possibility is well known from the theory of IF logic. In fact, (4.7) is equivalent to the IF sentence

$$(\forall x)(\forall z)(\exists y/\forall z)(\exists u/\forall x)S[x, y, z, u] \qquad (4.2)$$

while (8)–(9) are equivalent to the following ordinary first-order sentences:

$$(\forall x)(\exists y)(\forall z)(\exists u)S[x, y, z, u] \qquad (4.10)$$

$$(\forall z)(\exists u)(\forall x)(\exists y)S[x, y, z, u] \qquad (4.11)$$

As usual, the IF formula (4.2) is stronger than the corresponding ordinary first-order formulas (4.10)–(4.11). Thus (4.2) = (4.7) can fail to be true even though (4.10) = (4.8) and (4.11) = (4.9) are true. Moreover, it need not even be the case that (4.7) is false; it may simply fail to be true. (The reader will find more about this possibility in Chapter 7.)

Thus certain initially puzzling aspects of the conceptual situation in quantum theory, especially the impossibility of measuring conjugate variables at the same time, can be understood as examples of the characteristic behavior of quantifiers in IF logic. This fact, together with my earlier explanation of how naturally the distinctive features of IF logic emerge out of game-theoretical semantics, helps to demystify the indeterminacy phenomena of quantum theory.

There is another connection here with what is found in quantum theory. The formulas (4.8) = (4.10) and (4.9) = (4.11) can be taken to pertain to the two different orders of the two projective operations that characterize the measurement of x and z, respectively. The result is different in the two cases in that (4.8) = (4.10) and (4.9) = (4.11) are not equivalent. This amounts to the phenomenon of noncommutativity, which again turns out to be a manifestation of a simple feature of IF logic. How close we are here to the actual mathematics of quantum theory is shown by the well-known results (see, e.g., Hughes 1989, pp. 102–104) to the effect that compatible observables commute. For one can define compatible observables as two observables both of which depend functionally on one and the same third observable.

It is worth pointing out that in analyzing and explicating such notions as indeterminacy and failure of commutativity I have not appealed in the least to the physical situation in quantum theory, for instance to the alleged influence of the measurement apparatus on the object of measurement. I do not need such hypotheses here.

In mathematics, unlike physics, the informational independence of quantifiers ranging over different classes of individuals is sometimes taken for granted. As a result, an entire and extensive mathematical theory can be essentially a theory of certain kinds of informationally independent quantifiers. An important example is offered by what is known as the Ramsey Theory in combinatorics. It was launched by Frank Ramsey (1930) as a by-product of his work on the decision problem of first-order logic, and rediscovered in 1933 by a group of young mathematicians that included Paul Erdös and George Szekeres. The first problem that caught the attention of one of them (Esther Klein) can serve to illustrate the role of informationally independent quantifiers in Ramsey Theory. The problem was to prove the following theorem: If five points all lie in a plane but no three of them in a straight line, then four of the five always form a convex quadrilateral (cf. Graham and Spencer 1990, p. 114). The five original points can be thought of as being introduced by five universal quantifiers, say $(\forall x_1), (\forall x_2), (\forall x_3), (\forall x_4)$ and $(\forall x_5)$. The choice of the four can be formalized by an existential quantifier specifying the excluded point, say

$$(\exists y)((y = x_1 \lor y = x_2 \lor \cdots \lor y = x_5) \& \cdots) \tag{4.12}$$

The quantifiers by means of which it is stipulated that the other four form a convex quadrilateral clearly can – and must – be taken to be independent of the fifth and of $(\exists y)$.

Erdös and Klein quickly generalized the problem of $1 + 2^{k-2}$ given points and a convex k-sided polygon. In recalling the event, Szekeres subsequently wrote

We soon realized that simple-minded argument would not do and there was a feeling of excitement that a new type of geometrical problem emerged from our circle. (Graham and Spencer 1990, p. 114).

In the light of hindsight, we can say that the "new type of $\cdots$ problem" was a problem involving informationally independent quantifiers. This is the explanation of the nontriviality of proofs of relatively simple results in Ramsey Theory which was noted by Szekeres. In fact, the entire Ramsey Theory is shot through with informationally independent quantifiers.

This can be illustrated and documented by considering as an example one of the main results of Ramsey Theory, known as the Hales–Jewett Theorem. It can be formulated in the conventional terminology as follows:

For all r, t there exists $N' = HJ(r, t)$ so that for $N \geqslant N'$, the following holds: If the vertices of C_t^N are r-colored there exists a monochromatic line.

I have followed here the formulation of Graham, Rotschild, and Spencer (1990, pp. 32–35). One does not even have to know the precise definitions of the concepts used here to realize what is going on. (As an aid to visualization, C_t^N is roughly speaking an N-dimensional cube with edges of t units.) It is easily seen that the Hales–Jewett Theorem can be reformulated as follows:

JOYS OF INDEPENDENCE: SOME USES OF IF LOGIC

For all t, the following holds: For each R there exists N', and for each C_t^N together with a r-coloring of its vertices, there exists a line l in C_t^N such that, if $N \geqslant N'$ and $R \geqslant r$, then l is monochromatic.

Here "existing for" means being functionally dependent on. The point is that N' is a function only of t and R while l is a function of only t, N, and the r-coloring. For instance, for l we can take the line in C_t^N that has the largest number of elements of the same color. This brings into the open that there is a Henkin quantifier structure as part of the logical form of the Hales–Jewett Theorem.

Similar comments can be made of several other theorems of Ramsey Theory. As was mentioned, it seems to me that the overall character of Ramsey Theory is closely related to the pervasive use of informationally independent quantifiers in it. In fact, some of the peculiarities of informationally independent quantifiers have been noted precisely in the context of Ramsey Theory by logicians and linguists studying generalized quantifiers (see Keenan and Westerstahl, 1996). They have not pointed out the generality of the phenomenon in question, however.

The combinatorial complexity of IF first-order logic is related to certain types of computational complexity. This matter is studied in Hintikka and Sandu (1995). They consider a first-order arithmetical language L. With each formula F or L, they correlate a computer architecture capable of computing the Skolem functions of F (in the extended sense of Skolem function explained in Chapter 2). For an open formula $F[x]$ with, say, one free variable x, this computer architecture will then be able to compute for any given value x, whether this value satisfies $F[x]$ or not. For a closed formula (sentence) this computer can calculate the truth-value of the sentence in question.

This correlation between logical formulas and computer architectures is a generalization of the antediluvian or, rather, pre-von Neumann correlation between truth-functions and switching circuits. As was just presented, it is not complete when one goes beyond propositional logic in the sense that many consistent first-order sentences do not have recursive Skolem functions. No computer architecture is then correlated with them, notwithstanding their consistency. I will return to this matter in Chapter 10. It does not invalidate the use to which I am putting the (admittedly partial) correlation between first-order formulas and computing architectures.

How does this correlation behave? Suppose I am given two formulas. To each I can correlate a certain computer, but I can also combine the two formulas into a more complex one. This combined formula will be correlated (with a usually more complex) computer architecture.

This correlation can be extended to simple and more complex functional expressions (functional terms). Then the correlate of such a term will be simply a computer architecture that will compute the function in question.

Any such functional expression $f_0[x]$ can also be correlated with the formula

$$(\forall x)(\exists y)(y = f_0[x]) \qquad (4.13)$$

Clearly the Skolem function of (4.13) can be chosen to be $f_0[x]$. However, here we are considering formula–architecture correlations, and using the extension to functional expressions only as an expository device. Strictly speaking you should always think of all functions as having been eliminated from formulas I am discussing in terms of predicates in the usual way, rewriting, for example, $f(x) = y$ as $F(x, y)$ where it is assumed that

$$(\forall x)(\exists y)F(x, y) \qquad (4.14)$$

$$(\forall x)(\forall y)(\forall z)((F(x, y) \mathbin{\&} F(x, z)) \supset y = z) \qquad (4.15)$$

The failure of an analogy between first-order formulas and first-order functional expressions illustrates how easily one loses sight of informational independence by switching one's attention to functions. It is as if one had already opened the door a little bit to the axiom of choice merely by allowing function symbols into one's first-order language.

But suppose I start from two computer architectures, each correlated with a certain ordinary first-order formula. Furthermore, suppose that these computer architectures are then combined into a more complex one. Is there still an ordinary first-order formula correlated with the combined architecture? The right answer is: It depends. It is not hard to see that if the two given computers are combined sequentially, the result is a correlate of a more complex ordinary first-order formula. For instance, if the two formulas are functional terms $f_1[x]$ and $f_2[x]$, then the sequential combination corresponds to $f_2[f_1[x]]$.

It is easily seen, however, that when two computer architectures are combined in parallel (e.g., as components of a more comprehensive architecture) the result is no longer always a correlate of an ordinary first-order expression. For instance, consider the following combined architecture, where C_1, C_2, and C_0 are correlates of certain functional expressions $f_1[x]$, $f_2[x]$, and $f_0[x]$:

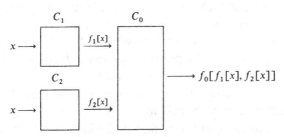

Thus there is a functional expression correlated with the combined architecture; but there is no longer an ordinary first-order formula correlated with it. For suppose C_1 and C_2 are correlated with

$$(\forall x)(\exists y)S_1[x, y] \quad \text{and} \tag{4.16}$$

$$(\forall z)(\exists u)S_1[z, u] \tag{4.17}$$

Then in the formula correlated with the combined architecture $(\exists y)$ must depend on $(\forall x)$ but not on $(\forall z)$. Likewise, $(\exists u)$ must depend on $(\forall z)$ but not on $(\forall x)$. Such a pattern of dependencies cannot be represented by any ordinary quantifier prefix.

However, it can obviously be expressed by means of the IF prefix

$$(\forall x)(\forall z)(\exists y/\forall z)(\exists u/\forall x) \tag{4.18}$$

Hence a parallel combination of architectures corresponds, not always to an ordinary first-order formula, but (always) to an IF first-order formula. In this interesting sense, *IF first-order logic is the logic of parallel processing.* For all practical purposes, it is such a logic in the same sense in which truth-function theory is the logic of switching circuits, albeit with the qualification to be registered in Chapter 10.

Independence-friendly logic can also be used to discuss matters of principle in the foundations of mathematics. Among other things, it throws additional interesting light on the axiom of choice. It was noted in the preceding chapter that in IF first-order logic we can have a complete disproof procedure, even though there is no complete proof procedure. Such a complete disproof procedure nevertheless cannot be obtained simply by extending the techniques of first-order reasoning to IF first-order logic. The usual rule of instantiation, especially the rule for existential instantiation, cannot do the whole job. The reason is intrinsic to the very ideas of quantifier dependence and independence. If we try to deal with quantifiers in IF first-order logic by means of successive instantiations, we lose track of which instantiations depend on which. We need a rule of functional instantiation for the purpose, as was indicated in the preceding chapter. The fact that it is possible to use only individual instantiation in ordinary first-order logic is a fortunate but atypical and accidental feature of this logic. In this respect, as in several others (cf. Chapter 2), ordinary first-order logic is simply too simple to serve as a paradigm case of what real logic is like in general.

In fact, we could use the rule of functional instantiation (cf. Chapter 3) also in ordinary first-order logic, and obtain an elegant proof method. Perhaps we ought to do so, not only because of the technical merits of the resulting proof method but primarily because it would bring out more perspicuously than the received methods the all-important relations of dependence and independence between different quantifiers. In any case, it is amply clear that the rule of functional instantiation formulated in Chapter 3 is an archetypally first-order (quantificational) rule. It is an immediate and natural expression of the normal interpretation of quantifiers as codifying choice functions of a certain kind (cf. here Hintikka and Sandu 1994).

If so, then the axiom of choice is vindicated, for, when applied to a sentence of the form

$$(\forall x)(\exists y)S[x, y] \tag{4.1}$$

where $S[x, y]$ is in a negation normal form with no existential quantifiers and no disjunctions, it yields as a consequence

$$(\forall x)S[x, f(x)] \tag{4.19}$$

where f is a new function constant. But such inferences are precisely what the axiom of choice justifies.

All this could – and should – be said with reference to ordinary first-order logic. What IF first-order logic does is, in the first place, to illustrate the organic place of the rule of functional instantiation in dealing with quantifiers of any kind by pointing out that in IF first-order logic some such rule is indispensable. The specific contribution that IF first-order logic makes here is to show that more complex and potentially more powerful variations of the axiom of choice can be justified in the same way. The inferences they authorize come about by applying the rule of functional instantiation to sentences more complex than (4.1).

Thus we obtain striking evidence not only of the acceptability of the axiom of choice but also of its character as a purely logical principle. Indeed, the rule of functional instantiation can be considered a generalization of the axiom of choice at the same time as it can be considered a generalization of the usual rule of existential instantiation in terms of individuals.

The most telling context nevertheless, in which the notion of informational independence plays a crucial role is, epistemic discourse, including the use of questions and answers. The notion of independence does not only offer a new and sharper tool for discussing epistemic logic and the logic and semantics of questions and answers. This notion is indispensable for the purpose of capturing the central concepts that figure in our use of such ideas as knowledge as well as question and answer. Here I will discuss only the logic of knowledge usually referred to as epistemic logic.

The role of informational independence in epistemic logic is not only important in its own right – it will later turn out to offer insights into the epistemology of mathematics (see Chapter 11). It is therefore in order to sketch briefly the role of informational independence in epistemic logic (for the concepts used here, see Hintikka 1992).

As usual, I take the notion of *knows that* as the sole primitive. The sentence *a knows that S* will be represented as $K_a S$. In addition to the notion of *knowing that*, we used the notion of epistemic possibility: $P_b S$ means *it is possible, for all that b knows, that S*. Since the only case considered here is one in which only one person's knowledge is being discussed, the subscript indicating the agent (knower) can sometimes be omitted.

It can be said that a statement like $K_a S$ deals with the *knowledge of facts* (or of *propositions*). The logic of such knowledge is relatively easy to master. I am taking here the usual semantics of such notions for granted. What is not equally clear is how our *knowledge of entities* (or of *objects*) of different kinds (other than facts, in case you think of facts as entities) can be understood and how it is related (to the extent it is related) to our knowledge of facts.

The most important kind of knowledge involving individuals, functions, and other entities (different from facts) is the kind of knowledge which in English is expressed by the different wh-constructions. How are they to be expressed in an explicit logical notation?

Earlier, attempts were made to deal with these constructions in terms of K plus ordinary first-order quantifiers. Several constructions can apparently be dealt with in such a manner, including the following:

a knows whether or not $S = K_aS \lor K_a \sim S$ (4.20)

a knows whether S_1 or $S_2 = K_aS_1 \lor K_aS_2$ (4.21)

a knows who satisfies the condition $S[x] = (\exists x)K_aS[x]$ (4.22)

But more complex kinds of knowledge cannot be so dealt with. Schematic examples illustrating this include the following:

a knows to whom (y) each individual (x) bears the relation
$S(x, y)$ (4.23)

a knows whether each individual (x) satisfies the condition
$S_1[x]$ or $S_2[x]$ (4.24)

It can be shown that the meaning of these sentences cannot be accounted for in terms of an interplay of K and first-order quantifiers (and propositional connectives). In terms of conventional logical notation, they can be expressed only by resorting to second-order quantification. Indeed, (4.23)–(4.24) are equivalent to the following:

$$(\exists f)K(\forall x)S(x, f(x)) \tag{4.25}$$

$$(\exists f)K(\forall x)((S_1[x] \ \& \ (f(x) = 0)) \lor (S_2[x] \ \& \ (f(x) \neq 0))) \tag{4.26}$$

But this poses additional problems. For one thing, there is no trace of higher-order quantification in the English sentences (4.23)–(4.24) any more than there is such a trace in (4.20)–(4.21). Why, then, must we use second-order quantifiers in (4.23) and (4.24)? And why should (4.23)–(4.24) have to be dealt with in an essentially different way from (4.21)–(4.22), when there is no clue to such a difference in these ordinary-language knowledge statements?

A method is found in this apparent madness when we realize that the English wh-elements, logically speaking, merely indicate independence of a sentence-initial K. This is true of such "propositional" questions as (4.20)–(4.21) which are usually not included among wh-questions as well as of such wh-questions proper as (4.22)–(4.23). (It is not even clear how (4.24) should be labeled in the received classification.) Thus (4.20)–(4.24) can be represented as follows:

$$K(S(\vee /K) \sim S) \tag{4.27}$$

$$K(S_1(\vee /K)S_2) \tag{4.28}$$

$$K(\exists x/K)S[x] \tag{4.29}$$

$$K(\forall x)(\exists y/K)S[x, y] \tag{4.30}$$

$$K(\forall x)(S_1[x](\vee /K)S_2[x]) \tag{4.31}$$

Here (4.25)–(4.26) are simply the second-order translations of (4.30)–(4.31). We can now see why (4.23)–(4.24) do not allow for a formulation in terms of a linear sequence of quantifiers plus K. If we try to express (4.30) on the first-order level without independent quantifiers, we run into an unsolvable dilemma. Since $(\exists y)$ depends on $(\forall x)$, it should come later than $(\forall x)$. But since it is independent of K, it should precede K and hence also $(\forall x)$. Similar remarks apply to (4.24) and (4.31).

The similarity between IF epistemic languages and English becomes even more pronounced when we use the double-slash notation. This is illustrated by the following translations:

a knows whether or not $S = (K_a// \vee)(S \vee \sim S)$ (4.32)

a knows whether S_1 or $S_2 = (K_a// \vee)(S_1 \vee S_2)$ (4.33)

a knows who (x) is such that $S[x] = (K_a//\exists x)(\exists x)S[x]$ (4.34)

a knows to whom (y) each individual (x) bears the
relation $S[x, y] = (K_a//\exists y)(\forall x)(\exists y)S[x, y]$ (4.35)

a knows whether each individual (x) satisfies the condition
$S_1[x]$ or $S_2[x] = (K_a// \vee)(\forall x)(S_1[x] \vee S_2[x])$

(4.36)

This is not the right occasion to develop further the theory of epistemic logic resulting from these ideas. There nevertheless is one massive fact that will later be relevant to my investigations in this book. In order to see what it is, we may consider the two epistemic statements

$$K(\exists x)S[x] \tag{4.37}$$

$$K(\exists x/K)S[x] \tag{4.29}$$

What does their truth mean model-theoretically? What (4.37) says is
that in each model (world, scenario) compatible with what is known
there is an individual x satisfying $S[x]$. In other words, it is known that
there is an x such that $S[x]$. What (4.29) says is that I can choose an
individual x, independently of whatever world my opponent chooses,
such that $S[x]$ is true in that world. This clearly means that it is known
who such an x is. The important semantical difference is that of these
two the former (4.37) depends only on what is true in each of the worlds
considered alone. No comparisons between different worlds are made.
The latter, that is (4.29), depends also on what counts as the same
individual in different worlds. In order to specify the semantics of
knows that statements, the former is enough. In order to specify the
semantics of *knows* + wh-constructions we also need criteria of cross-
world identity. They are not fixed by what is true in each world taken
alone.

In brief, the criteria of *knowing that* do not completely determine
what counts as *knowledge of individuals* and other kinds of entities,
for instance what counts as knowing who, what, when, where, and so
forth. This result is eminently in agreement with what is found in
ordinary discourse. In order to understand, for instance, a *knows who*
statement, one must know what the criteria of identification are that
the speaker or writer is presupposing. In so far as we are dealing with
objective meaning, these criteria are objective. Moreover, the same
criteria must be used throughout any coherent argument or discourse. But they do not reduce to the criteria of *knowing that*
(knowing facts). Accordingly, different criteria can be adopted, and
are adopted, on different occasions. *Knowing who someone is* means
somewhat different things to passport authorities, the FBI, or to the
registrar of your university; and it means radically different things to
the occult writer who once asked in a *Boston Globe* ad: "Who were
you in 1493?" and to the anonymous British male chauvinist pig who
coined the saying, "Be nice to young girls. You never know who they
will be." A choice between all these different criteria of identification
is determined not by some eternal and immutable logic, just because
the underlying logic will remain the same structurally no matter

which choice is made; the choice is rather guided by general epistemological considerations plus the specific purpose of the discourse in question.

This observation can be extended to knowledge of higher-order entities, for instance, knowledge of functions. This is an important point, for functions are inevitably involved as soon as an existential quantifier depends on a universal one. For instance, in order to be in a position to say that a scientist s knows that a variable y depends on another, say x, in a certain way specified by $S[x, y]$ it does not suffice that there exists a function f such that he knows that

$$K_s(\forall x)S[x, f(x)] \qquad\qquad (4.38)$$

This function f must furthermore be known to s. Only then does (4.38) imply

$$K_s(\forall x)(\exists y/K_s)S[x, y] \qquad\qquad (4.39)$$

And the fact that f is known to s is expressed by

$$K_s(\forall x)(\exists y/K_s)(f(x) = y) \qquad\qquad (4.40)$$

Now the truth-condition of statements like (4.39) are not determined by the truth-conditions of unslashed statements of the form K_sS. In order to apply epistemic concepts to functions, we must therefore have independent criteria as to when they are known.

5

The Complexities of Completeness

In view of what was said in Chapters 3–4, there is no doubt about the character of IF first-order logic as the basic area of our logic. But what is this logic like? What is new about it? In Chapters 3–4, a number of partial answers were given to these questions. Perhaps the most general one concerns the expressive strength of our new basic logic. In view of the close similarity between ordinary and IF first-order logic, it may come as a surprise that IF first-order logic is much stronger than its more restricted traditional version. How much stronger will become clearer in the course of my examination of its properties and applications.

One consequence of the strength of IF first-order logic is that it does not admit of a complete axiomatization. The set of valid formulas of IF first-order logic is not recursively enumerable. Hence there is no finite (or recursive) set of axioms from which all valid sentences of this logic can be derived as theorems by means of completely formal (recursive) rules of inference. Thus the first remarkable property of IF first-order logic is that, unlike its special case of ordinary first-order logic, it does not admit of a complete axiomatization. The reasons for this incompleteness will be explained in Chapter 7.

How are we to react to this incompleteness? Different perspectives are possible here. Purely technically, in view of the power of IF first-order logic, the failure of this logic to be axiomatizable is perhaps not entirely surprising. Yet this failure (and this power) goes against a philosophical tradition of long standing. According to this traditional way of thinking, incompleteness is a problem for mathematicians, not for logicians. It was in many ways shocking of Gödel to prove that a mathematical theory as simple as elementary arithmetic is incomplete. Yet the nearly universal assumption has been that this incompleteness is somehow attributable to the character of elementary arithmetic as a mathematical theory. Logicians and philosophers could find some consolation in the thought that pure logic, which typically was taken to be first-order logic, remained untainted by the dreaded virus of incompleteness.

Admittedly, certain areas of what is commonly called logic, for instance second-order logic, have been known to be incomplete. But the typical reaction to their incompleteness has been to reclassify them as mathematical rather than purely logical theories, perhaps amounting to set theory (or a part thereof) considered as a mathematical theory.

This kind of reaction to Gödel's incompleteness result is not a peculiarity of twentieth-century philosophers. It is in reality part and parcel of a long tradition which goes back at least to Descartes. Descartes rejected traditional syllogistic logic as being trivial and fruitless, while extolling the power of mathematical reasoning, especially of the analytic method of Greek geometers.[1] In a similar vein, Kant declared logic as being based on the law of contradiction and yielding only analytic truths, in contradistinction to the synthetic a priori truths of mathematics. Against such a background, it was not surprising that Gödelian incompleteness was thought of as being due to the status of elementary arithmetic as a mathematical theory.

Yet a closer analysis of the history of this tradition soon produces warning signals. Whatever the right interpretation of the ancient analytic method is, and whatever one thinks of its use in the hands of Descartes and his contemporaries, it is a fact that it was used by Greek geometers in the context of reasoning which can be captured by the resources of our ordinary first-order logic. Furthermore, for Kant the gist of the mathematical method consisted in the use of instantiation rules, which with our twentieth-century hindsight are again part and parcel of logic in our sense, that is, of first-order logic. Hence it is clear that at the earlier stages of this tradition the boundary between logic and mathematics was drawn in an essentially different way from what twentieth-century philosophers are used to. Yet this historical discrepancy has not really been acknowledged by most philosophers. They go on thinking in terms of an essential difference between unproblematic logic and treacherous mathematics. This may, among other things, explain why Hilbert's ideas have not been appreciated more widely. One thing that he tried to do was to break down the traditional dichotomy between logical and mathematical reasoning.

Against this background, it is a shock indeed that the most basic part of our logic, IF first-order logic, is incomplete. This shock is not just a matter of being unused to the idea that a genuine part of logic could be incomplete. For one thing, it forces us to look at the entire history of the foundations of mathematics in a new light. For instance, suppose, as a light-hearted thought-experiment, that Frege had corrected his fallacy, admitted informationally independent quantifiers and realized that the resulting logic is not axiomatizable. This would have made nonsense of his entire project. For what would then have been the point of reducing mathematics to logic, if the logic needed for that purpose cannot be axiomatized? How could Frege have possibly been able to characterize the logic which the reduction aims at, since he avoided semantical methods and used only formal (syntactical) ones? Again, Hilbert could not have conceived of his famous project of proving semantical (model-theoretical) consistency of mathematical theories by establishing their proof-theoretical consistency, if he had realized that the true basic logic he needed in those theories would not have a complete proof-theoretical axiomatization.

It is also instructive here to recall what was found in Chapter 2. It was pointed out there that, in a perspective provided by GTS, ordinary first-order logic is in many respects unrepresentative of the conceptual situation in general. One of the reasons why IF first-order logic is interesting is that it is more representative of the whole range of things that can happen in logic in general than ordinary first-order logic. It represents a concrete example of features that are characteristic of logic in general but are not exhibited by its more traditional cousins. Incompleteness is one such feature. Some others will be examined in the next couple of chapters.

In order to understand the conceptual situation created by the incompleteness of IF first-order logic, we have to also take into account Gödel's celebrated impossibility results (1931). I do not think it is any exaggeration to say that the community of philosophers and mathematicians has still not managed to cope with the true import of Gödel's discoveries and to draw the right consequences from them for the future development of the foundations of mathematics.

First, we have to realize that there are several entirely different notions of completeness. They do not even apply to the same sort of thing. For instance, what Gödel was dealing with was the incompleteness of a nonlogical (mathematical) theory, namely, elementary arithmetic, whereas the incompleteness of IF first-order logic pertains to a part of logic, namely, to the nonaxiomatizability of a logical theory. It is a serious mistake to assimilate these two kinds of theories to each other. Such an assimilation in effect means overlooking the distinction between the two functions of logic in mathematics which was emphasized in Chapter 1.

We have to distinguish from each other, at the very least, the following senses of completeness:

(a) *Descriptive completeness.* It is a property of a nonlogical axiom system T. It means that the models of T comprise only the intended models. If there is only one intended model (*modulo* isomorphism), descriptive completeness means categoricity, with the intended model as the sole model of T.

Notice that no reference to any method of logical proof is made in this characterization. All that is needed is the relation of a sentence to its models plus the idea of an intended model.

If, following the terminology of logicians, the set of intended models is said to constitute a theory, descriptive completeness means that this theory is axiomatizable. This is precisely the sense in which Euclid and Hilbert were trying to axiomatize geometry. It is also the sense in which we speak of the axiomatization of some physical theory. In that case, the intended models are simply all the physically possible systems of the relevant kind.

(b) *Semantical completeness.* It is a property of a so-called axiomatization of (some part of) logic. It means that all the valid sentences of the underlying language can be obtained as theorems from the so-called axioms of that system of logic by means of its inference rules. If we assume the usual permissive standards of axiomatization, then there exists a complete axiomatization for some part of logic if and only if the set of logically true (valid) sentences (of that part of logic) is recursively enumerable.

(c) *Deductive completeness* is a property of a nonlogical axiom system T *together with* an axiomatization of logic (method of formal logical proof). It means that from T one can prove by means of the underlying logic C or $\sim C$ for each sentence C of the language in question.

(d) There is a further use of the term "complete" in connection with nonlogical axiom systems. It can be called *Hilbertian completeness*, for the *locus classicus* of this sense of the term is the so-called axiom of completeness first used in the second edition of Hilbert's *Grundlagen der Geometrie* (1899, second ed. 1902). Strictly speaking, though, the axiom occurs first in Hilbert (1900). The difficulty of understanding the notion of completeness is illustrated by the confusions surrounding Hilbert's "unglückseliges Axiom" (Freudenthal 1957, p. 117). The sense of completeness it involves was first assimilated to the other sense of completeness. Only in Baldus (1928) do we find a clear statement of the difference between Hilbertian completeness and other senses of completeness.

Hilbert's axiom is in effect a maximality assumption. It says that the intended models of the axiom systems are such that no new geometrical objects can be added to them without violating the (other) axioms of the system. Maximality assumptions of this kind play a potentially important role in the foundations of mathematics but the maximality of a model with respect to the individuals it contains is a fundamentally different idea from the maximality of the system, which is what the descriptive and deductive senses of completeness deal with.

The difference between these different senses of completeness will become clearer as we proceed. Philosophers may have been confused here by the fact that there are interrelations between these different kinds of completeness and incompleteness. For instance, let us assume that there is only one intended model for a nonlogical theory T which is based on a semantically complete logic L. Then T is descriptively complete only if it is deductively complete. However, if L is not semantically complete, then T can be descriptively complete even when it is deductively incomplete.

The differences between (a)–(c) are not merely technical. They are manifestations of the different purposes which logic can serve in mathematical thinking. These purposes were discussed in Chapter 1. Completeness in the sense (a) pertains to the descriptive function of logic in mathematics which was described there. If a mathematician has reached a descriptively complete axiom system, then he or she has reached an intellectual mastery of the range of structures that are exemplified by the intended models of *T*. The mathematician has separated these structures from all the others and expressed what is characteristic of them.

In contrast to such descriptive completeness (a), deductive completeness (c) deals with something else. It deals with the deductive function of logic, which in this case amounts to a mathematician's ability to deal computationally with the given theory *T*. Completeness in this sense means that there is an algorithm for separating what is true and what is false in what can be said of the models of *T*. Such an algorithm takes the form of a set of rules for deriving theorems from axioms. An emphasis on this kind of completeness (deductive completeness) often reflects a conception according to which the essence of mathematical practice is theorem-proving. It also means emphasizing the deductive function of logic at the expense of its descriptive function.

The general distinction of which the contrast between descriptive completeness and deductive completeness is a special case is not new. As an example, logicians almost routinely make a distinction between descriptive complexity and computational complexity. The former deals with the power of a logic to capture different kinds of structures "out there", while the latter deals with the complexity of the different derivations of theorems from axioms. This obviously parallels the distinction between descriptive completeness on the one hand and deductive (as well as semantical) completeness on the other (see, e.g., Dawar and Hella forthcoming).

What are the main consequences of Gödel's incompleteness theorem? What did he really prove? First and foremost, it is important to realize that the only kind of incompleteness that he established directly is the deductive incompleteness of elementary arithmetic. That is, there is no recursive enumeration of the true sentences of elementary arithmetic implemented by an axiomatization of the underlying logic. In other words, no consistent axiomatization *T* of elementary logic, taken together with an explicitly axiomatized logic *L*, enables us to prove (i.e. to prove from *T* by means of the logic *L*) either *S* or $\sim S$ for each sentence *S* expressible in the language of elementary arithmetic.

As was just noted, from this deductive incompleteness one can infer the inevitability of the descriptive incompleteness of elementary arithmetic only if *L* is semantically complete. This was the case with the logic Gödel actually used. He used ordinary first-order logic, which he had proved semantically complete himself in Gödel (1930). But the observations made in this work put the matter in a new light. It has been seen that the completeness of ordinary first-order logic is merely a by-product of the unnecessary and unnatural restrictions that Frege and Russell imposed on the formation rules of this logic in ruling out independence-friendliness. Hence it is likely that we have to use a semantically incomplete logic anyway. And if so, Gödel's results do not imply anything about the prospects of formulating descriptively complete axiom systems for elementary arithmetic, not even on the usual first-order level.

Here we can see one kind of impact that IF first-order logic has in virtue of its very existence. One implicit reason why the possibility of formulating descriptively complete nonlogical theories based on a semantically complete logic has not been taken seriously has been noted previously. It is the belief that no logic worth its name can be incomplete. IF first-order logic provides a concrete counterexample to such beliefs, even though it does not itself provide the possibility of a descriptively complete axiomatization of elementary arithmetic. In fact, it will be shown later in this work how such semantically complete axiomatizations can in fact be formulated by methods suggested by the game-theoretical approach to logic and mathematics. In this way, the developments reported in this book put Gödel's incompleteness results in a new and interesting perspective.

From this perspective, it must be said that the philosophical community has not acknowledged how limited the significance of Gödel's incompleteness result really is. This point is related to what was said earlier of the difference in the significance of the different kinds of completeness, especially the difference between descriptive and deductive completeness. This misperception has deeper roots. Philosophers have been impressed by Gödel's result because they have overestimated the importance of deductive and computational techniques in mathematics. They have been seduced by the oversimplified picture of mathematical activity as mere theorem-proving. In reality, it is clear that deductive incompleteness is not the most important kind of incompleteness. What attracted mathematicians, scientists and philosophers in the first place to the axiomatic method was the possibility of an intellectual mastery of a whole area of important truths, say all geometrical truths. A complete axiom system was a means of reaching that mastery. But this kind of mastery is in the last analysis a mastery of the models of the axioms, and not of the things we (or our computers) can deductively prove about them. Such mastery can manifest itself in ways other than a mechanical derivation of theorems from axioms. Contrary to the oversimplified picture that most philosophers have of mathematical practice, much of what a mathematician actually does is not to derive theorems from axioms. This point is especially clear in those branches of mathematics which do in fact start from a given set of axioms, for instance, the theory of groups. What a group theorist or a mathematician working in a similar field does is only to a limited extent theorem-proving in the strict sense of the term. Most of it can from a logical point of view be characterized only as a metatheory of group theory. It is a quest for an overview of all the models of a given theory, and not of its deductive consequences. A mathematician for instance classifies groups and proves representation theorems for different kinds of groups. Such theorizing is largely independent of the deductive resources which the mathematician in question commands.

The kind of completeness relevant to this kind of enterprise is obviously *descriptive* completeness. This notion is definable completely independently of any axiomatization of the underlying logic, and hence independently of all questions of deductive completeness. It admittedly presupposes that we understand the sentence–model relationship. But this we need to master in order to understand the meanings of our sentences in the first place. It does not presuppose, as we just saw, that the axiom system in question is deductively complete.

Hence Gödel's incompleteness result does not touch directly on the most important sense of completeness and incompleteness, namely, descriptive completeness and incompleteness. This is an important observation, for Gödel's result has generally been taken to point out limits to what can be done by means of logic. The explanation of this discrepancy in perception is that the distinction between the different kinds of completeness has not been made clearly by the majority of philosophers and mathematicians. This failure has even led to the panic reaction of doubting even the viability of the concept of arithmetical truth. In reality, Gödel's incompleteness result casts absolutely no shadow on the notion of truth. All that it says is that the whole set of arithmetical truths cannot be listed, one by one, by a Turing machine.

Thus we have reached an interesting insight into Gödel's incompleteness theorem. This result pertains to the deductive and computational ways of dealing with nonlogical axiom systems. Noting this true address of Gödel's result helps us to understand the urgency of the theoretical problems which Gödel's incompleteness theorem has forced us to face. In many ways these problems have become increasingly more pressing by the developments which have occurred since Gödel reached his theorem. In the meantime, computational and deductive ways of dealing with problems of virtually any kind have become increasingly more common in almost all walks of life because of the explosive development of computer technology. Such computational techniques are prevalent in areas like cognitive science and artificial intelligence. It follows from Gödel's and Tarski's results (or is at least strongly suggested by them) that computational techniques can never exhaust the theory of the phenomena in question. Hence the status of most of the computational approaches to different kinds of inquiry are bound to stay forever in a limbo of incompleteness. This seems to deprive computational methods of all deeper theoretical interest. The entire development of computer-based technology for dealing with language and thinking is thus not only under suspicion but already indicted for being ad hoc, a mere collection of tricks and rules of thumb put together for the purposes of application. Such overall suspicions are of course unjustified. However, philosophically speaking a justification presupposes a shift of emphasis away from attempts to capture arithmetical truths, or whatever phenomena one may be studying, in one fell swoop computationally to the never-ending pursuit of better and better approximations of the target phenomena.

Thus in a sense Gödel's real message was merely the deductive incompleteness of elementary arithmetic. As such, his famous result affects only the prospects of our deductive mastery of elementary arithmetic and not our ability to deal with this branch of mathematics axiomatically or descriptively, unless such axiomatic treatments are constrained to use only ordinary first-order logic.

Admittedly, Gödel's incompleteness results are closely related to the undefinability of arithmetical truth in elementary arithmetic itself. However, this aspect of Gödel's results can be judged in the same way as Tarski's result (1935, 1956a) concerning the undefinability of truth for a language in that language itself, and will be so judged in Chapter 6.

Meanwhile, a few comments of another kind may be in order. A confusion between deductive and semantical completeness (and

COMPLEXITIES OF COMPLETENESS

incompleteness) is also what underlies the fashionable comparisons between non-Euclidean geometries and nonclassical logics. The possibility of non-Euclidean geometries is a consequence of the descriptive incompleteness of an axiomatic geometry without the parallel postulate. But different kinds of logics are not models of any one incomplete theory that could be completed by adding new axioms. In order to obtain different logics, it is not enough to choose different axioms. For this purpose, you have to define different notions of model and/or truth. You cannot as much as speak of the semantical completeness of a logical axiomatization without having defined the notions of model and truth and, by their means, the notion of validity.

The incompleteness that attaches to IF first-order logic is naturally semantical incompleteness. As such, it has few implications for the completeness or incompleteness of nonlogical theories. On the contrary, the enhanced strength of IF first-order logic makes it *ceteris paribus* a better tool for formulating descriptively complete nonlogical theories, notwithstanding its semantical incompleteness.

Indeed, as soon as we have a viable notion of truth-in-a-model at our disposal, and also some characterization of the totality of all models of the given language, then we can set up an axiom set for elementary arithmetic hoping realistically that it might then turn out to be complete. The intended kind of completeness – I have called it descriptive completeness – will then be different from deductive completeness. For one thing, nothing is said in the characterization of this kind of completeness of any actual deductive method; the question of deductive completeness does not even have to be raised in order to be able to speak of descriptive completeness. Descriptive completeness is in a sense a purely model-theoretical notion.

Unfortunately this implies that descriptive completeness is subject to "Tarski's curse" discussed in Chapter 1. The notion of validity, that is, truth in every model, cannot be characterized without characterizing the notion of truth; which in turn cannot be defined (if Tarski's result is the last word on the subject) without resorting to a stronger metalanguage, for instance, to higher-order logic or set theory. And that appeal to set theory seems to entail all the problems, puzzles and paradoxes that set theory is ripe with.

One thing that I will do in this book is to study different ways out of this problem. For instance, as far as Gödel's immediate subject-matter, elementary arithmetic, is concerned, descriptive completeness would be easy to reach if we were allowed to ascend to the second-order level. Essentially, all we have to do is to replace in a Peano-type axiom system the schema for induction by the second-order formulation of the principle of induction:

$$(\forall X)((X(0) \ \& \ (\forall y)(X(y) \supset X(y+1))) \supset (\forall y)X(y)) \qquad (5.1)$$

Of course, the standard interpretation in Henkin's (1950) sense of the second-order language being used must be presupposed. Unfortunately, as was pointed out in Chapter 1, second-order logic apparently involves all the vexing problems of set existence, and is therefore deemed by many philosophers to be beyond the pale of pure logic.

And even on the first-order level, there are ways of obtaining descriptively complete axiomatizations of elementary arithmetic. It follows from Gödel's result that such axiomatizations must not use ordinary first-order logic or any other deductively complete logic. But it is not difficult to find other kinds of first-order logic to fit the bill. Indeed, in Chapter 7 it will be shown how elementary arithmetic can be completely axiomatized (in the descriptive sense of completeness) on the first-order level. Another, and somewhat more exotic, way of doing so will be briefly discussed in Chapter 10.

There are other ways of illustrating the fact that Gödel's results do not cover all the bases even in the foundations of arithmetic. For instance, even if elementary arithmetic were deductively complete, it would not yet be descriptively complete. In more literal terms, even the deductively complete theory which is true in the structure of natural numbers is not descriptively complete.[2] (Needless to say, this theory is not axiomatizable.) It admits of several different non-isomorphic models; it does not have the structure formed by natural numbers (with addition and multiplication) as its only model. Indeed, the descriptive incompleteness of Peano arithmetic is an almost direct consequence of the compactness of first-order logic. It does not need Gödel's elaborate argument for its proof. All this shows how many important issues are left untouched by Gödel's results.

In a different direction, it is known that there are deductively complete and hence decidable theories of reals and of sufficiently elementary geometry (see, e.g., Tarski 1951, 1959). These theories are nevertheless – or, for that very reason – descriptively incomplete. They admit of different nonisomorphic models, and hence will not satisfy a mathematician who like Hilbert is committed first and foremost to the study of certain kinds of structures, such as the structure of the "real" reals. Such a mathematician will not be impressed by the deductive completeness of a Tarski-type theory of reals.

One upshot of this incomplete examination of the problems of incompleteness is a virtual reversal of the traditional picture sketched above. Incompleteness is a logical rather than a mathematical phenomenon. There are in principle no obstacles for mathematicians to reach the kind of completeness that really matters to them, namely descriptive completeness. In subsequent chapters, I will try to say a little bit more of the concrete ways of realizing such completeness. But in order to reach such completeness, the underlying logic must be strengthened so as to become semantically incomplete.

In this respect, IF first-order logic is tailor-made to satisfy the existing need of such a strong and natural, but at the same time, semantically incomplete logic.

These brief remarks enable us to describe the general problem situation that Gödel's incompleteness result forces on us. The crucial feature of the situation is not that we cannot somehow set up descriptively complete systems of axioms for the mathematical theories we are interested in – I will soon return to the question as to how we can actually set them up. The possibility of descriptively complete axiomatizations nevertheless does not diminish the importance of the problem of finding actual deductive ways of dealing with the structures that the axiom system deals with. Hence the main post-Gödelian, not to say postmodern, foundational problem is to look for new deductive methods and to analyze them. Gödel's results show that this cannot be done mechanically, much less all at once, but of course this is precisely the reason why this new "fundamental problem of the philosophy of mathematics" is an intriguing one. In practice, such stronger aids of deduction can often be codified in the form of new axioms for the mathematical theory in question. Hence the task I am talking about here is not entirely unlike the task of finding stronger and stronger axioms of set theory.

Here the impact of Gödel's incompleteness theorem is reinforced by that of the incompleteness of IF first-order logic. In the preceding chapter, it was seen that mathematics is shot through with concepts that rely on IF first-order logic. What this means is that the full logic that is needed in most of the advanced mathematical theories is not amenable to a semantically complete axiomatization. Mathematical reasoning, for instance mathematical theorem-proving, cannot be reduced to the deduction of new results by means of a logic that can be completely captured by any set of recursive rules. Moreover, the new principles of reasoning whose need is always lurking just under the surface are in principle but logical rather than set-theoretical.

For this reason, ordinary first-order axiomatic set theory is an inappropriate framework for mathematical theorizing when it comes to finding new and stronger principles of mathematical reasoning, even if the need for finding ever stronger set-theoretical assumptions is acknowledged.

More generally speaking, as soon as we realize that the fundamental logic of mathematics is IF first-order logic rather than its old-fashioned brother, we lose any and every reason to formulate deductively complete theories in mathematics, except for atypical special cases. For an axiomatic theory can be deductively complete only if the underlying logic is semantically complete when restricted to deductions from the axioms of that theory. And usually this is possible only if that basic logic is semantically complete in general.

For instance, one can now state concisely one reason why set theory is inappropriate as a basis of mathematics. The problem is not that set theory has led to paradoxes or threatens to lead to paradoxes. The problem is not the deductive incompleteness of set theory, either. It is its descriptive incompleteness. This is the reason why set theory cannot provide aims, much less guidelines, for a search of increasingly stronger new assumptions which can be thought of as providing better deductive methods for dealing with it.

This descriptive incompleteness pertains in the first place to first-order axiomatizations of set theory. But in its usual forms axiomatic set theory does not have really natural higher-order models, either. I will return to the prospects of axiomatic set theory in Chapter 8.

So what should a mathematician do? When a logician or a mathematician sets up an axiomatic system these days his or her main aim invariably is to make it deductively complete. I can envisage a future in which the typical aim of a mathematician or logician is to formulate descriptively complete axiom systems which are deductively incomplete. This is the kind of axiom system that can best serve the crucial purpose of guiding the mathematicians' and logicians' never-ending quest for increasingly stronger deductive principles for that particular theory. Descriptive completeness provides then the aim which can help to direct the logicians' search for stronger logical principles that would provide more deductive power.

Such descriptively complete axiom systems must nonetheless be on the first-order level. For otherwise they will be subject to Tarski's curse: their model theory would arguably involve set-theoretical concepts.

This observation provides an answer to a question that may already have occurred to the reader. Is not the task of finding new axioms addressed every day by working logicians, at least tacitly? For the obvious suggestion here is that guidelines for the search for new deductive principles are provided by the model theory for that part of mathematics in question. For instance, in elementary arithmetic we have a nice truth definition which can guide a logician's or mathematician's search for stronger axioms.

Somewhat similar things can be said of a logician's quest for semantically complete axiomatizations of some part of logic. This kind of completeness is, as I have shown, a mirage aready on the level of IF first-order logic. What can be done, however, is to give a non-axiomatic characterization of logical truth (validity) and to use it as a guideline for looking for stronger deductive principles. In practice, it is difficult (and pointless) to try to distinguish the quest for stronger deductive principles sharply from the quest for stronger descriptive axioms.

But this cannot be the whole story. For one thing, not all model theory helps us in our search for the right deductive principles. For the purpose, we must have some idea of what the intended models of the relevant theory are so that we can characterize truths in those models. Furthermore, we must gain some new insights into the structure of the intended models in order to see what follows logically from the (descriptively complete) axioms characterizing those structures. Thus it is not the whole story that deductive logic is a means of understanding better the structures a mathematical axiom system specifies. Sometimes, the traffic flows in the other direction: better grasp of those structures is needed to see what the deductive consequences of the axioms are.

Admittedly, this "reverse logic" situation does not seem to arise very often in mathematical practice. But perhaps this rarity is something of an illusion. For instance, when number-theoretical results are proved by analytical means what happens is that the structure of natural numbers is considered as part of a more complicated structure, insights into which provide insights also into the structure of natural numbers. Another example is provided by the way in which the continuity assumptions concerning real numbers were originally inspired by the analogy between the structure of reals and the structure of a geometrical line.

A search for stronger deductive assumptions is also closely related to that fetish of recent philosophy of mathematics, namely, mathematical practice. As was pointed out above, a mathematician is typically searching for an insight into the structure of all models of a given theory. Only a minuscule part of a practising mathematician's actual work consists in deductively deriving theorems from axioms, philosophers' preconceptions notwithstanding. Admittedly, such representation theorems do not usually give rise to new axioms for the original given theory or to new logical principles. Nevertheless, when the given theory is deductively incomplete, such meta-theorems may be one way of finding new axioms. I am even tempted to suggest that the role Gödel assigned to intuition in the foundations of mathematics should really be understood as pointing out the importance of this metatheoretical and model-theoretical way of deriving better deductive principles for the study of certain kinds of structures from insights concerning these structures directly.

I will return to the problem of finding stronger assumptions in Chapters 8–9. Further reasons will be given there as to why axiomatic set theory serves singularly badly the new purposes of foundational research that Gödel's incompleteness results force logicians to face.

These viewpoints can be illustrated in terms of the very distinction made earlier in this chapter between different senses of completeness. Doubts are perhaps prompted by my characterization of descriptive completeness, the reason being the reference it contains to the class of intended models. These intended models are sometimes referred to also as standard models. However, in this usage the term "standard" is merely a euphemism for "intended", for it cannot in general be identified with any of the senses of standardness (such as Henkin's in his 1950 paper) that can be strictly defined. It is not even clear what the relationship between these different senses of standardness is, or may be.

Hence any reference to intended models in the definition of descriptive completeness might seem extrinsic to the true foundations of mathematics, and perhaps even dangerously unclear. How can we characterize such models with sight unseen, so to speak? One way of allaying such doubts is to point out that the idea of the intended model need not be any more dubious than the idea of the model in general. As was in effect pointed out in Chapter 1, the general concept of model also involves prelogical ideas concerning what a structure must be like in order to be acceptable as a model *simpliciter*. One way of seeing this is to examine how the notion of model can naturally be varied. Such variation helps us to understand what the unvaried notion of model involves.

One can for instance, widen the concepts of model and obtain a weaker logic as a consequence – that is, a logic in which fewer formulas are valid because there are more models for them to fail to be true in. A way of doing so is easily uncovered by means of GTS. It shows that our usual logic is based on an important assumption. The different choices made by the two players can be thought of in the same way as probability theorists think of their sampling procedures, to wit, as "draws" (selections) of "balls" (individuals) from an "urn" (domain of a model). Normally, it is assumed that the population of balls in an urn stays the same between successive draws in the logical terminology; that the model does not change between the successive moves in a semantical game. This assumption is not unavoidable. Probability theorists have in fact devised "urn models" which are changed in accordance with some fixed law between the draws. In the same way, we can allow the model on which a semantical game is based to change in certain specified ways between the successive moves of the game. The result is a well-defined logic which can be used in the study of the foundations of mathematics (see Hintikka 1975, Rantala 1975).

In the other direction, one can change one's logic by imposing extra conditions on it. This will result in there being more logical truths than before, for there are now fewer acceptable models in which a formula can fail to be true. Once again, the resulting new logic can have interesting uses in the study of the foundations of mathematics. An especially interesting possibility in this direction is to implement maximality and minimality assumptions concerning models, not by means of explicit axioms or other kinds of sentences, but as a model of the underlying language in the first place. This possibility is tentatively explored in Hintikka (1993a).

Implicit in the traditional notion of a model is a rejection of changes of both these two kinds. Hence this traditional notion of model involves highly nontrivial prelogical assumptions. It is not any more of a dangerous assumption to think that somehow a narrower class of intended models of some mathematical theory is given to us. Certainly we have as sharp an idea of the structure of natural numbers as we have of traditional models as distinguished from their rivals.

It seems to me that the usefulness of our prelogical ideas about certain classes of structures could be exploited more widely than they have in fact been utilized. For instance, the second number class (the structure of countable ordinals) is in our day usually studied as an aspect of axiomatic set theory. I would not be surprised if it turned out to be helpful to draft into the service of some set theorists in a more direct way the intuitions we have about the second number class, by considering its model theory directly in its own right.

We have thus seen that the rumors of the demise of completeness in mathematics have been greatly exaggerated. The most important kind of completeness in mathematics is alive and well in the land of model theory. However, the independence of that land still appears to remain in jeopardy, as was emphasized in Chapter 1. For its basic notion, the notion of truth in a model, still seems to us to take refuge in higher-order considerations which appear to invite an invasion of higher-order logical (or else set-theoretical) foreign armies. My next task hence is to see how this danger can be avoided.

Notes

[1] For the history of the analytic method of ancient Greek geometers and for its subsequent history, cf. Hintikka and Remes (1974). For Kant, see Hintikka (1973).

[2] In this sense, a "theory" is simply a deductively closed set of formulas.

6

Who's Afraid of Alfred Tarski?
Truth Definitions for IF First-Order Languages

In earlier chapters, it was seen how crucial the concept of truth is in the foundations of logic and mathematics. An especially central role was seen to be played by the question of the definability of truth. The new languages explained in Chapter 3, IF first-order languages, open up a way of actually freeing the expressibility of the concept of truth from the serious problems that have been seen to beset it.

As was pointed out in Chapter 1, discussions of truth and its definability have in the last sixty years been profoundly influenced by the ideas of Alfred Tarski (1935, 1956a).[1] This influence has been strong and widespread. The following aspects of Tarski's work are especially relevant here:

(i) In his classic monograph, Tarski showed how the concept of truth can be defined explicitly and precisely for certain types of formal languages.

    The nature of Tarski-type truth definitions was explained briefly in Chapter 1. Among other things, the following features of these definitions were pointed out:

(ii) A Tarski-type definition of truth is an indirect one, defining the notion of truth in tandem with the notion of satisfaction.

(iii) Tarski proved, given certain assumptions, that a definition of truth for a formal language can only be formulated in a metalanguage which is stronger in certain specifiable respects than the given object language. In our day, Tarski-type truth predicates for first-order languages are usually expressed by Σ_1^1 second-order formulas – that is to say, by formulas whose only non–first-

order ingredient is a formula-initial string of second-order existential quantifiers.

(iv) Tarski argued that the concept of truth cannot be defined for
what he called the colloquial language.

(v) A less deep but eminently popular suggestion of Tarski's was to
use as a touchstone for truth definitions that they have as their
consequence all the relevant instances of the so-called T-schema:

(T) Π is true $\leftrightarrow p$

where "Π" is a placeholder for a quote or a structural description
of the sentence which is to replace "p."

It will be argued in this chapter that we have to revise our ideas
about all the five features (i)–(v) of Tarski-type truth definitions.
It will also be seen that truth definitions can be freed from the
problems – or at least limitations – which they were seen in
Chapter 1 to be subject to.

Tarski's procedure is generally taken to be normative. However, it
is seldom, if ever, explained what the reason for this normative
character of Tarski's definition is, or even what Tarski's own rationale was in proceeding in the way he did. There is nevertheless a
simple answer available, which is the same for both questions. This
answer is closely related to the presuppositions of the approach
used here.

Historically, an important part of Tarski's background was an
early form of the idea of categorial grammar. Now any approach like
categorial grammar presupposes certain things of the semantics of
the language whose grammar it is supposed to be. It presupposes
that the language in question satisfies the requirement which linguists usually call *compositionality*. (For this notion, see Partee 1984
and Pelletier 1994.) Philosophers sometimes refer to it as the Frege
principle, and Frege did indeed pay lip service to it. As usually
formulated, it says that the meaning of a complex expression is
a function of the meanings of its constituent expressions. It is
nevertheless appropriate to extend the idea of compositionality to
say that all the relevant semantical attributes of an expression (and
not only just its meaning or truth) are functions of the semantical
attributes of its constituent expressions (not just of its meaning). The
major impact of the principle of compositionality on logical and
linguistic theorizing is to allow what is usually referred to as

recursive definitions (or other kinds of recursive characterization) of the relevant semantical properties. That is to say, we can specify the conditions for the applicability of the semantical attribute in question to the simplest possible expressions, and then specify, step by step, for each operation of forming complex expressions out of the simpler ones, how applicability of that semantical attribute to the complex expression is derived on the semantical attributes of its constituent expressions.

This idea of compositionality has loomed large in recent linguistic and philosophical theorizing. Conformity with the principle of compositionality is generally (though not universally) thought of as a major desideratum which a satisfactory linguistic theory should satisfy. Donald Davidson apparently at one time even thought of compositionality as a precondition for the learnability of language (see Davidson 1965).

The real impact of the principle of compositionality nevertheless has not been emphasized in recent discussion. It is illustrated by the role of the principle in facilitating recursive definitions of semantical attributes. Such recursive definitions proceed from simpler expressions to more complicated ones. They are not possible unless the attribute to be defined is semantically speaking *context-independent*. The main function of the principle of compositionality is to secure such semantical context-independence.

It seems to me that the principle of compositionality was one of the most important presuppositions of Tarski's work on the concept of truth. For one thing, it was fairly obvious that it was the principle of compositionality that motivated Tarski's reliance on recursive definitions in his approach to truth. Furthermore, it is the same commitment to inside-out procedure that forced Tarski to define truth for his formal languages, not alone, but in tandem with the notion of satisfaction. In a compositional procedure, one defines the semantical attributes of an expression in terms of the semantical attributes of its component expressions. But truth can only be attributed to sentences (closed formulas), whose component expressions are often not sentences but open formulas. For this reason, truth can be defined compositionally only in conjunction with some other semantical concept like satisfaction. This observation helps us to understand Tarski's procedure in his classic paper. It also offers a glimpse of the difficulties in which adherence to the principle of compositionality easily leads a linguistically minded logician or logically minded linguist. For clearly it would be preferable to define the notion of truth directly without using the concept of satisfaction as a middleman. For one thing, we are likely to have sharper pretheoretical ideas about the notion of truth than about the notion of satisfaction.

Tarski points out himself the roots of his ideas in the notion of semantical category which goes back of Leśniewski and indirectly to Husserl (cf. Tarski 1956a, p. 215). As is spelled out in various explicit developments of categorial grammar, the leading idea here is to mirror the semantical structure of a sentence by the way in which it is constructed syntactically from its basic components. Whether or not such parallelism between semantics and syntax inextricably presupposes compositionality, there is little doubt that in praising the naturalness of the notion of semantical category Tarski is in effect extolling the virtues of compositionality.

In short, adherence to the principle of compositionality is the answer, or at least a large part of the answer, to the question as to why Tarski formulated his truth definition in the way he did. Furthermore, I suspect that the same principle was operative in Tarski's claim that truth is not definable for our ordinary "colloquial" language. Tarski blamed this alleged impossibility on the irregularities of natural languages. I suspect that a failure to conform to the principle of compositionality was the first and foremost "irregularity" that Tarski discerned in natural languages.

It is not that prima facie counterexamples to compositionality have not been noted in the literature. The nonsynonymy of the following pairs of sentences is a case in point:

Mary will be surprised if anyone comes.   (6.1)

Mary will be surprised if everyone comes.   (6.2)

Likewise, the context affects the meaning of a sentence like

Jim can beat anyone.   (6.3)

when it is imbedded in a belief sentence:

John doesn't believe that Jim can beat anyone.   (6.4)

For what John is said to disbelieve in (6.4) is not (6.3) but

Jim can beat someone.   (6.5)

It is even possible to give a systematic account of the reason why compositionality fails in sentences like (6.1) or (6.4). An important explanation for one type of counterexample lies in what I have called the special ordering principles of GTS. They do not respect the order in which an expression is assembled out of its component expressions (see here Hintikka and Kulas 1983, pp. 233–234; 1985, pp. 180–181). In spite of all this, linguists have not taken prima facie counterexamples to compositionality as seriously as they ought to have done. Maybe they have never read those chapters of the treatises in the philosophy of science that deal with the significance of anomalies in heralding theory changes.

One of the most important kinds of impacts that IF first-order logic has on our ideas about logic and language, is that it shows once and for all the utter futility of trying to abide by the principle of compositionality in our linguistic and logical theorizing. The reason stares you in the face from the explanations that were given in Chapter 3 of the nature of IF logic. By their very nature, all such instances of quantifier independence (of any other kind of informational independence in logic) as cannot be dealt with in ordinary first-order logic violate the principle of compositionality.

This could be dramatized by changing the notation that has been used here along the lines mentioned in Chapter 3. Instead of appending to each quantifier an indication of which earlier quantifiers it is independent of, we could add to each quantifier a list of later quantifiers that are exempted from its scope. For instance, instead of writing my favorite sample formula in the notation so far used as

$$(\forall x)(\forall z)(\exists y/\exists x)(\exists u/\forall z)S[x, y, z, u] \tag{6.6}$$

I could write it as

$$(\forall x//\exists u)(\forall z//\exists y)(\exists y)(\exists u)S[x, y, z, u] \tag{6.7}$$

where the double slash // is the converse of the independence relation /.

I have argued in print that compositionality is a lost cause in the study of the semantics of natural languages (see Hintikka and Kulas 1983, Ch. 10). This is not an impossibility claim. If you are a die-hard believer in compositionality, there is a way in which you can always enforce it. The prima facie violations of compositionality are due to the semantical interaction of an expression with its context, as we have seen. Now in principle you can try to build the laws covering that interaction into the semantical entity you consider as the valu

110 THE PRINCIPLES OF MATHEMATICS REVISITED

of the expression in question. In practice, however, this would result in intolerably complicated and completely unnatural semantics.

This unnaturalness is illustrated by the fact that a similar act of semantical desperation is impossible in suitable formal languages. There the resurrection of compositionality is made impossible by the tacit conventions that govern the interpretation of logical formulas. No perverse ingenuity can make the semantics of sentences like (6.6)–(6.7) conform to compositionality.

The commitment of Tarski-type truth definitions to compositionality is so deep that the glory and misery of the entire principle of compositionality is to a large extent measured by the success and shortcomings of Tarski-type truth definitions. Since I am in this chapter showing the very important limitations of Tarski-type definitions, I am ipso facto exposing certain serious theoretical shortcomings of the principle of compositionality.

Conversely, to look at the bright side of things, the very existence of IF first-order logic is an eloquent proof that a rejection of compositionality is no obstacle to the formulation of a simple and powerful logic. Indeed, the best argument against compositionality as a general linguistic principle is the success of independence-friendly logic in its different variants in the logical analysis of various important concepts and of their manifestations in natural languages. This is not the occasion to tell that success story, which is in fact still continuing. A cumulative evidence of different applications is in any case sufficiently impressive for me to rest my case against compositionality for the purposes of this book (see here, e.g., Hintikka and Kulas 1983 and 1985, Hintikka and Sandu 1991 and 1996).

The dependence of Tarski-type truth definitions on the assumption of compositionality makes them suspect even when they do work.

Because IF first-order languages are not compositional, Tarski-type truth definitions do not apply to them. You just cannot give satisfaction-conditions for expressions like

$$(\exists y)(\exists u)S[x, y, z, u] \tag{6.8}$$

when you do not know what kinds of dependence and independence relations obtain there between $(\exists y)$, $(\exists u)$ and the quantifiers further out to which x and z might be bound (cf. 6.7).

This failure of Tarski-type truth definitions in IF first-order logic is nevertheless perhaps less obvious than it might at first seem. It is at its clearest when we use the double-slash notation. For then there is no difference between two component expressions whose variables are in one of them independent of certain quantifiers further out, while in the other they are bound to such quantifiers. For instance, the universal-quantifier-free constituent expressions of the following two expressions are identical:

$$(\forall x)(\forall z)(\exists y)(\exists u)S[x, y, z, u] \tag{6.9}$$

$$(\forall x//\exists u)(\forall z//\exists y)(\exists y)(\exists u)S[x, y, z, u] \tag{6.7}$$

If compositionality were in force, the same semantical entity should be assigned to

$$(\exists y)(\exists u)S[x, y, z, u] \tag{6.8}$$

in the two cases. But this would make it impossible to distinguish (6.9) and (6.7) semantically if we abide compositionally and build our truth-conditions and satisfaction-conditions for (6.9) and (6.7) from inside out. For the "insides" of (6.9) and (6.7) are identical, and by the time we come to the slashed quantifiers it is too late to try to distinguish the two from each other semantically.

When the single-slash notation is used, we can make a distinction etween the semantical entities assigned, for example, to

$$(\exists u)S[x, y, z, u] \tag{6.10}$$

and

$$(\exists u/\forall x)S[x, y, z, u] \tag{6.11}$$

But this possibility will not help compositionalists decisively. As a minor point, compositionality will not be true in the literal sense of the word, for (6.11) is not a component expression of anything as it is not a well-formed formula. Whether or not it can occur as a part of a well-formed sentence depends on that part of the sentence which lies outside (6.11).

This failure of compositionality might seem to be mere quibbling. It is nevertheless indicative of the situation. In any case, there is another reason for the impossibility of Tarski-type truth definitions for IF first-order sentences. The difference between (6.10) and (6.11) lies in the question whether the value of u depends on the value of x. No matter how this dependence or independence is expressed in assigning a semantical value to (6.11), by the same token this assignment should indicate whether the value of u depends on those of y and z. But this question does not depend on the expression (6.11) alone, as compositionality requires. Because of the convention that was shown in Chapter 3 to be needed in IF first-order languages, an answer to the question depends also on whether y and z are bound to universal or existential quantifiers. And this cannot be seen from the component expression in question alone. Likewise, it cannot be seen from (6.11) as to which of the variables x, y, z depend on which ones. Yet this is needed if we want to assign a class of valuations to (6.11) as those that satisfy it.

As a consequence, there is no realistic hope of formulating compositional truth-conditions for IF first-order sentences, even though I have not given a strict impossibility proof to that effect.

This failure of Tarski-type truth definitions is an extremely serious black mark against them. Most of the earlier failures of such definitions took place at the outer reaches of ongoing logical research and involved concepts that are at least prima facie unfamiliar – unless you are Leibniz and relish the kind of infinite analysis that infinitely deep languages codify. In contrast, IF logic was seen to be our true elementary logic involving nothing more than quantifiers and propositional connectives. Hence the failure of Tarski-type truth definitions in IF first-order logic shows that they do not have a serious claim to being the basic normative kind of truth definition.

In contrast to Tarski-type truth definitions, game-theoretical characterizations of truth are easily available for IF first-order languages. This is not much of a surprise, for a semantical game starts from an entire sentence and not from its simplest constituents, and gradually anatomizes it into simpler and simpler sentences. In such an outside-in process, context-dependencies can easily be taken into account.

Indeed, for each first-order sentence S (whether IF or not) its second-order translation can serve as its truth-condition. For, as we saw in Chapter 3, such a "translation" expresses precisely the existence of a winning strategy for the initial verifier in the game $G(S)$ correlated with S. This is how truth is characterized in the first place in GTS. The remaining part of the task of formulating an actual truth *definition* is in effect to integrate all the truth-conditions into a single definition of truth. It will turn out, Alfred Tarski notwithstanding, that such a truth definition can be given for representative IF first-order language in that language itself.

Let us approach such truth definitions step by step. I will first discuss the definability of arithmetical truth. There the model in which the truth or falsity of the sentences of the underlying language is evaluated is the structure of natural numbers. First, I will likewise assume that the language that is being used is an ordinary first-order arithmetical language. Later, I will expand my horizon to take in independence-friendly quantifiers and connectives also, plus the Σ_1^1 fragment of the corresponding second-order language.

I assume that my readers are familiar with the technique of Gödel numbering. The basic idea is completely straightforward. What is involved is a coding of the syntax of the given formal object language in question in the language of elementary arithmetic. For the purpose of this discussion, it will be assumed that the object language itself contains some suitable formalization of elementary (first-order) arithmetic. The reason is to show that for certain object languages the truth predicate can be defined in the object language itself.

The coding of the syntax of the object language by means of the arithmetical part of itself can be done in different ways. All that counts here is that certain formal relations between formulas can be expressed in the language itself. In my metalogical discussion, I will refer to the Gödel number of a formula S by means of corner quotes. In other words, $\ulcorner S \urcorner$ is the Gödel number of S. The numeral which represents a natural number n will be referred to as **n**.

Then the properties of, and relationships between, different expressions whose representability plays a role in my line of thought are relations between expressions of the following forms:

    (i) n and **n**   (6.12)

    (ii) $(S_1 \,\&\, S_2)$, S_1 and S_2

    (iii) $(S_1 \vee S_2)$, S_1 and S_2

    (iv) $(\forall x)S[x]$, n and $S[\mathbf{n}]$

    (v) $(\exists x)S[x]$, n and $S[\mathbf{n}]$

If $R(x, y)$ is a primitive two-place relation,

    (vi) $R(\mathbf{n}, \mathbf{m})$, n, m

Likewise for other primitive predicates and primitive functions.

    (vii) Likewise for negated primitive predicates and functions.

    (viii) S_1 , S_2 when S_2 is the negation normal form of S_1

This relation will be called $R(x, y)$.

Later, the Gödel numbering technique will be also applied to the Σ_1^1 part of the corresponding second-order language with independence-friendly quantifiers allowed. (It will be assumed for simplicity that all the initial second-order existential quantifiers are function quantifiers.) Then we will consider relations between the following:

(ix) $\ulcorner (\forall y)(\forall u)(\exists z/\forall u)(\exists t/\forall y)S[y, z, u, t)]\urcorner$, k, l, m, n and $\ulcorner S[k, l, m, n]\urcorner$

How do I know that all these relations are representable in elementary number theory and hence in the given IF first-order language? A blow-by-blow account of an argument to this effect would be too cumbersome to be given here. Fortunately, there is an argument which is not only persuasive but resorted to by logicians every day. It is an appeal to Church's thesis. It is known that all recursive relations are representable in elementary number theory (cf. e.g., Mendelson 1987, p. 143, Proposition 3.23). Church's thesis says that every effectively (mechanically) decidable relation is recursive. Since all the relations discussed here are fairly obviously mechanically decidable, they are representable in the given IF first-order language, for it was assumed to contain elementary arithmetic. In some cases, the representability of these relations in a suitable language of arithmetic is known from the literature.

For the purposes of a perspicuous exposition, I will not introduce any special notation for the relations (i)–(ix) but use the corner quote notation instead.

Once we have all the relations (i)–(ix) expressible in the language, it is easy to see how a truth predicate can be formulated for different languages. I will first show how a truth predicate for an ordinary first-order language L can be formulated in the corresponding IF first-order language. I will first formulate this predicate in the Σ_1^1 fragment of the corresponding second-order languages. Once this is done, one can simply translate that predicate back to the corresponding IF first-order language. By a corresponding language, I mean of course a language with the same nonlogical primitives.

The second-order truth predicate has an obvious intuitive meaning. It says, when applied to the Gödel number y of a sentence, that there exists a one-place predicate X which behaves in the way a truth predicate should, and that y has this predicate. In other words, the truth predicate has the form

$$(\exists X)(\text{TR}[X] \ \& \ X(y)) \tag{6.13}$$

Here $\mathrm{TR}[X]$ serves to guarantee that there is a winning strategy for the initial verifier in a semantical game. In fact, it is a conjunction of the following formulas:

(a) $(\forall x)(\forall y)(\forall z)((x = \ulcorner (S_1 \& S_2)\urcorner \& y = \ulcorner S_1 \urcorner \& z = \ulcorner S_2 \urcorner) \supset (X(x) \supset (X(y) \& X(z))))$

(b) likewise for disjunction

(c) $(\forall y)(\forall z)(\forall u)((y = \ulcorner (\forall x)S[x]\urcorner \& u = \ulcorner S[z]\urcorner \& X(y)) \supset X(u))$

(d) $(\forall y)(\forall z)(\forall u)((y = \ulcorner (\exists x)S[x]\urcorner \& u = \ulcorner S[z]\urcorner \& X(y)) \supset X(u))$

(e) If R is a two-place primitive relation, $(\forall x)(\forall y)(X(\ulcorner R(x,y)\urcorner \leftrightarrow R(x,y))$; likewise for other primitive predicates and negated primitive predicates.

(f) $(\forall x)(\forall y)((x = \ulcorner S_1 \urcorner \& y = \ulcorner S_2 \urcorner \& N(x,y)) \supset (X(x) \leftrightarrow X(y)))$
where $N(x,y)$ is the relation of the Gödel number of a sentence to the Gödel number of its negation normal form

Once you understand this definition of the truth predicate (6.13), you will realize that it does its job. For the property of being true (as attributed to the Gödel numbers of sentences of a given ordinary first-order language with finite number of primitive predicates and functions) satisfies $\mathrm{TR}[X]$. Hence if the sentence with the Gödel number y is true, it satisfies (6.13).

Conversely, if it satisfies (6.13), from $\mathrm{TR}[X]$ you can see that there is a winning strategy for the initial verifier. That player can always make his or her choices in such a way that the Gödel number of the resulting sentence has the property X. Indeed, the truth predicate (6.13) is little more than a way of spelling out more fully the game-theoretical truth definition explained in Chapter 2.

This observation can be formulated in a somewhat different way. Consider, for the purpose, the statement

$$(\exists X)(\mathrm{TR}[X] \& X(\ulcorner S \urcorner)) \tag{6.14}$$

in which the truth predicate is applied to the Gödel number of the sentence S. It is easily seen that (6.14) is logically equivalent to the game-theoretical (second-order) truth-condition of S, as explained in Chapter 2. This helps to explain the sense in which the truth predicate (6.13) merely serves to integrate the game-theoretical truth-conditions for different sentences.

The most remarkable thing about the truth predicate (6.13) is that it is of the Σ_1^1 form. (The only higher-order quantifier in (6.13) is the initial second-order existential quantifier $(\exists X)$.) From this it follows that it can be translated into the corresponding IF first-order language[2]. Thus a truth predicate for a given ordinary first-order language can be defined in the corresponding IF first-order language; in other words without any higher-order quantifiers. Or, in other words, without having to raise any questions about the existence of higher-order entities.

How can we generalize the truth predicate (6.13) to other first-order languages besides purely arithmetical ones? It turns out that the natural way of doing so involves going back to the familiar Tarski-type procedure of first defining satisfaction for the language in question. The definition can be formulated in the corresponding second-order language. The satisfaction relation $\text{Sat}(x, f)$ holds between the Gödel number $x = \ulcorner S \urcorner$ of a formula S and a valuation function f if and only if f satisfies S in the sense explained in Chapter 1. As was also pointed out in Chapter 1 the valuation function f is a mapping from natural numbers to the individuals of the model in question. I assume here and in the following that the given first-order language contains elementary arithmetic. Then the valuation function f is just an ordinary second-order object, and the existentially quantified formula

$$(\exists f)\text{Sat}(\ulcorner S \urcorner, f) \tag{6.15}$$

is simply a Σ_1^1 formula of the corresponding second-order language. When S is a closed formula (i.e., a sentence), (6.15) asserts that S is true.

So why did I offer a different truth predicate (6.13) for arithmetical languages? My reasons were pedagogical. First, the direct truth predicate shows vividly how extremely close the notion of truth is to both our ordinary ideas about truth and to the basic ideas of the game-theoretical approach represented in this book.

Second, I wanted to bring out the fact that the difficulty of extending the truth predicate (6.13) to other first-order languages besides arithmetical ones is in a sense merely technical, and has nothing to do with the deep problems of compositionality or of the allegedly unavoidable second-order character of truth definitions for first-order languages. The feature of (6.13) that makes a generalization difficult is the quantification over numerals (names of numbers) in clauses (c) and (d) of the truth predicate. In order to extend such a truth predicate to nonarithmetical languages, one needs to have a name in the language for each individual in the models of the theory in question, and one needs even to be able to quantify over the class of all such names, just as in elementary arithmetic we have a numeral representing each number and can quantify over the set of all such numerals.

No matter whether we use the truth predicate (6.13) or (6.15), it is of Σ_1^1 form, and hence can be translated to the corresponding IF *first-order* language. In other words, either truth predicate is expressed without any quantification over higher-order entities. They are purely nominalistic (combinatorial), and free from any set-theoretical commitments. This is the first step in my exorcism of Tarski's curse. Other advantages ensue from my definition not presupposing compositionality. They will be discussed later in this chapter.

The truth predicate (6.13) does not work in IF first-order languages. However, it is easily modified so as to do so. Consider for the purpose an arithmetical IF first-order language. In such a language, we can define a pair predicate in the usual way. Now it has been shown by Krynicki (1993) that in an IF first-order language with a pair predicate all independent quantifiers can be expressed in terms of the familiar quantifier combination

$$(\forall x)(\forall z)(\exists y/\forall z)(\exists u/\forall x) \tag{6.16}$$

known as the Henkin quantifier. Moreover, the elimination of other independent quantifiers in favor of Henkin quantifiers can be taken to be effective. Hence we can strengthen the normal form predicate $N(x, y)$ as follows: It will now say that $y = \ulcorner S_2 \urcorner$ is the Gödel number of a normal form of the sentence with the Gödel number $x = \ulcorner S_1 \urcorner$, where S_2 satisfies the following requirements:

(i) S_2 is logically equivalent with S_1.
(ii) S_2 is obtained from S_1 by a recursive procedure.
(iii) S_2 is in the negation normal form.
(iv) In S_2, each slashed quantifier occurs in a Henkin quantifier (6.16).
(v) All slashed disjunctions are eliminated from S_2.

After this redefinition of $N(x, y)$, clause (f) in the definition of $TR(X)$ can read as it did before. The main novelty that is needed is an additional clause for Henkin quantifier formulas. It can be formulated as follows:

$$(\exists x)((x = \ulcorner(\forall y)(\forall u)(\exists z/\forall u)(\exists t/\forall y)S[y, z, u, t]\urcorner \& X(x)) \supset$$

$$(\exists f)(\exists g)(\forall y)(\forall u)(\exists w)(w = \ulcorner S[y, f(y), u, g(u)]\urcorner \& X(x)) \qquad (*)$$

A moment's thought shows that (∗) does indeed yield the right truth-condition for Henkin quantifier sentences. With the changes indicated, (6.13) serves as a truth predicate for the arithmetical IF first-order language in question.

The main difference is that (6.13) does not any longer contain only one initial second-order quantifier $(\exists X)$. Second-order quantifiers can also be introduced by (g), as shown by the quantifiers $(\exists f)$ and $(\exists g)$ there. But when these quantifiers are moved to an initial (prenex) position in (6.13), they still remain existential. Hence (6.13) as a whole is of a Σ_1^1 form. Hence (6.13) can be translated back into the original IF first-order language. All told, it follows that the truth predicate for the given model of an IF first-order language is *expressible in the very same IF first-order language.*

This truth predicate ((6.13) modified) does the job it was supposed to do. It provides a truth definition for a suitable IF first-order language in that language itself. Of course, the term "definition" has to be dealt with gingerly here. What has been shown is that there is a complex predicate (of the Gödel numbers of the sentences of the given IF first-order language) which applies to a number if and only if it is the Gödel number of a true sentence. Nothing is said of falsity here, and no explicit formal definitions are set up.

Like the earlier truth definition for ordinary first-order languages, the new one can be generalized from arithmetical languages to a large class of other IF first-order theories. This class includes all first-order theories which contain elementary arithmetic and also contain a pair predicate for all their individuals. In view of Krynicki's (1993) result, the latter assumption again implies that all slashed expressions occur in Henkin quantifiers.

Hence the only novelty needed here (as compared to a corresponding definition for ordinary first-order theories) is a clause in the characterization of satisfaction which takes care of Henkin quantifier formulas, in rough analogy to the clause (∗) given above. This is a straightforward matter. All we need to do is to say that a valuation v satisfies (6.6) if and only if there are functions $f(x)$ and

$g(z)$ such that for every x and z, $S[x, y, z, u]$ is satisfied by the valuation which is like v except that the value of y is $f(x)$ and the value of u is $g(z)$. This turns my satisfaction predicate $\text{Sat}(x, v)$ itself into Σ_1^1 form. But this fact does not mean that the truth predicate (6.15) is not also in Σ_1^1 form.

The reliance of my truth predicate (6.13) on existential function quantifiers brings the truth definition presented here even closer to the explanations given in Chapter 2 as to what truth means in GTS. In fact, the most remarkable thing about the truth predicate (6.13) is its close relationship to semantical games, especially to the initial verifier's strategies in these games. Indeed the entire definition is little more than an explicit formulation of the idea that the truth of a sentence S means the existence of a winning strategy for the initial verifier in the corresponding semantical game. The clauses for propositional connectives say in effect that the verifier must always be able to make a move connected with them in accordance with a winning strategy, in order for the propositional compound to be true. And the clause for quantifiers likewise says that the verifier of a true sentence must be able to choose in quantifier moves in accordance with a winning strategy if the quantified sentence is to be true. Because of the failure of compositionality in IF languages, this requirement must sometimes be made for several quantifiers at the same time (as in the Henkin quantifier), and not for each quantifier one by one.

It can in fact be shown that the sentence in which the truth predicate (6.13) is applied to the Gödel number of some particular sentence S is equivalent to the second-order sentence which states the game-theoretical truth condition of S.

Similar things can be said of the more general truth predicate (6.15), even though the detour via the notion of satisfaction tends to obscure them somewhat. It is for the purpose of avoiding this un-necessary partial obfuscation that I once again formulate my truth predicate first for IF first-order arithmetical languages, where the nature of my truth predicate can be appreciated most easily.

Since semantical games are essentially activities of verification and falsification, what all this means is that my truth predicate ties the concept of truth essentially to the activities by means of which the sentences of the relevant language are verified and falsified. The fact that these activities are not what in everyday life (and in everyday science) is most frequently meant by the verification, falsification, confirmation, or disconfirmation of propositions does not invalidate my point. The reasons for the basic position of semantical games in the constitution (to use the Husserlian term) of the concept of truth were explained in Chapter 2. With the qualifications expressed there, my truth definition can literally be taken to say that a sentence is true if and only if it can in principle be verified, that is, if and only if there exists a winning strategy for the initial verifier. Admittedly, the remarks just made relate directly to the second-order (Σ_1^1) formulation of my truth predicate. But the IF first-order truth definition is simply a translation of this predicate into the corresponding IF first-order language.

It follows that my truth predicate refutes an important (and fashionable) objection to explicit formal truth definitions. This objection is the main reason why Tarski-type truth definitions are these days routinely claimed to be philosophically irrelevant. It is sometimes put by saying that Tarski-type truth definitions merely characterize a certain abstract relationship between sentences and reality, but that they do not provide any reasons why this relationship should be taken to be the notion of *truth*. In another formulation, it has been alleged that what is wrong with Tarski-type truth definitions is that they are completely unrelated to the activites by means of which we actually establish the truth (or falsity) of different sentences. A Wittgensteinian might continue this line of thought and aver that without such a link between truth definitions and the activities of verification and falsification, it does not make sense to speak of teaching, learning, understanding or mastering the concept of truth so defined. As one recent writer has put it, Tarski-type truth definitions do not show "what it is in virtue of which a sentence has any truth-condition at all or in virtue of which it has the particular truth-condition it happens to have" (Cummins 1989, p. 11).

Whether or not these criticisms are in fact applicable to Tarski-type truth definitions need not be discussed here. The crucial thing is that the truth definition and the truth-conditions explained in this work answer precisely those questions which Tarski-type definitions have been claiming to be incapable of answering. For instance, why does a quantificational sentence S have the truth-condition it in fact has? Now what does the truth-condition say? It says that there exists a winning strategy for the initial verifier in the corresponding game $G(S)$. But how is the game $G(S)$ determined, besides of course the model (world) **M** with respect to which it is played? Its several moves and their order are determined by the syntactical structure of S. In this way, S has the truth-condition it has in virtue of its syntactical structure, but only indirectly, because of the way this structure determines the structure of the verification "game" $G(S)$.

In Chapter 10 I will show how the game-theoretical truth definition and game-theoretical truth-conditions open the door for further pertinent questions concerning the strategies that are their pivotal element; to wit, whether those strategies may perhaps be subject to some specifiable limitations.

More generally speaking, my truth definition does precisely what Tarski-type truth definitions were accused of not doing. They relate the notion of truth directly to the activities (semantical games) by means of which we in a sense verify and falsify sentences. Even the qualifications which are implicit in the phrase "in a sense" and which were explained in Chapter 2, do not invalidate my point.

Indeed, we have a near-paradox on our hands. Emphasis on an inseparable connection between truth and the activities of demonstrating truth is a characteristic of the constructivistic approach to the foundations of logic and mathematics. Hence we seem to have taken sides with the constructivists. Yet the outcome is an extension of classical logic capable of serving classical mathematics. This near-paradox occasions a fresh look at constructivistic ideas which will be undertaken in Chapter 10.

The independence of my truth definition, and indeed of the entire GTS, of the assumption of compositionality is not merely a philosophical or architectonic virtue. It extends the range of applicability of truth definitions. It was explained earlier why their commitment to compositionality prevents Tarski-type truth definitions from being applicable to IF first-order languages. This fact has been tacitly acknowledged by logicians in that special case of IF first-order languages, that is, languages with partially ordered quantifier prefixes, which they have studied since 1960 or so. In fact, all truth-conditions that have been offered for them are in explicitly game-theoretical terms, albeit formulated independently of my GTS (see Henkin 1961 and cf. e.g., Barwise 1979, Enderton 1970 and Walkoe 1970).

But this is not the only advantage offered by GTS and by my truth predicate. Since Tarski-type truth definitions toe the line of compositionality, they must operate from inside out in a given sentence. Hence they need starting-points for the definition in the form of unanalyzable atomic sentences. In contrast, game-theoretical truth definitions of the kind considered in this work operate from outside in. The game starts from the given sentence, and in each move it is replaced by a simpler sentence as an input for the next move. Now, as every game theorist knows, there is in principle no obstacle to defining winning and losing strategies, and so forth, also for infinitely deep languages. Hence GTS and my truth definition can in principle be applied to such Leibnitian languages as allow for infinitely deep formal expressions. This possibility has been put into practice by the group of logicians in Helsinki led by Jouko Väänänen and Juha Oikkonen who have systematically studied such infinitely deep languages. Needless to say, all their truth-conditions have been game-theoretical, as are the truth definitions used in the special case of infinitely deep languages known as game quantifier languages, whose study antedates the more general study of infinitely deep languages.[3]

Thus working logicians have repeatedly found Tarski-type approaches inapplicable in different circumstances, and invariably they have spontaneously resorted instead to game-theoretical conceptualizations.

The game-theoretical approach to truth and truth predicates has also distinct advantages over the approaches which operate with sequences of partial truth predicates and which were initiated by Kripke (1975). In an appendix to this book, Gabriel Sandu presents a detailed analysis of some of these advantages.

Furthermore, a question that may have bothered my readers receives a simple answer through a recognition of the basic role of the strategic viewpoint. The truth definition presented here is as clear as you can ever hope to get when presented in its second-order form. Yet its translation into IF first-order logic results in a complex and clumsy sentence which I have not even dared to try to write down explicitly. Similar things can be said of the truth-conditions of particular first-order sentences. Why this intuitive preference for the second-order formulation?

An answer is implicit in the original preformal game-theoretical characterization of truth. This characterization is in terms of the existence of winning strategies for the initial verifier. These strategies are expressed by (finite sets of) functions. Hence the most direct ("intuitive") formulation of game-theoretical truth-conditions is in second-order (Σ_1^1) terms, and one's reasoning about first-order truth is clear and *übersichtlich* only as long as it is likewise conducted on the second-order level.

However, this does not make any substantial difference to the systematic situation. A Σ_1^1 language can in principle be thought of as a notational shorthand alternative for the corresponding IF first-order language. At worst I can defend myself by appealing to something like Aristotle's distinction between what is primary for us humans and what is primary in the order of nature – or in any case, in the order of logic.

At the same time, we can see that there is a solid systematic reason for developing truth-theory in practice on the second-order level. For it is on that level that we have most immediate access to strategies for the players in semantical games.

To return to the starting point of this chapter, Tarski's approach to truth definitions, we have seen that Tarski-type truth definitions can, and must, be replaced by much more flexible definitions as do not presuppose compositionality (cf. item (i) in the initial list).

As to (ii), the usual detour via the notion of satisfaction in formulating the truth predicate is needed only because in most nonarithmetical languages there is no name for each member of the domain of individuals of the model in which truth is being defined.

As to the crucial claim (iii), it has been seen that the assumptions of Tarski's impossibility theorem are not satisfied by IF first-order languages and that it is in fact possible to define a truth predicate for a suitable IF first-order language in that language itself.

In Chapter 7 it will be seen that the T-schema is not appropriate as an adequacy criterion for truth definitions in general.

This leaves open only question (iv), that is, the definability of truth in ordinary ("colloquial") language. At first sight, I seem to be vulnerable to the same objection as has been repeatedly leveled at Tarski. From the fact that truth is definable for certain formalized (but interpreted) languages in the language itself it does not follow that truth is likewise definable for our actual working language. However, it is interesting to see that the evidential situation concerning the definability of truth for natural languages is radically different vis-à-vis my truth definitions from what it was for Tarski-type ones. As was pointed out, Tarski's doubts concerning the definability of truth for our "colloquial language" were probably prompted by the failure of compositionality in natural languages. In my treatment, such a failure becomes a reason for, instead of a reason against, expecting that truth is definable for natural languages. For dispensing with compositionality was precisely what enabled us to extend the scope of our truth definitions beyond the purview of Tarski-type ones.

Again, I argued ages ago (Hintikka 1974) that informationally independent quantifiers occur in natural languages. Some of my examples raised eyebrows initially, but the general upshot of my arguments is by now generally accepted. What has not been equally generally acknowledged yet is that different variations of informational independence play a ubiquitous role in the semantics of natural languages. They turn out to be at the bottom of such diverse semantical phenomena as the *Sprachlogik* of our epistemic vocabulary, of the *de dicto* versus *de re* distinction, and of the so-called nonstandard quantifiers. In the next chapter, it will be argued that negation behaves in natural languages very much in the same way as it does in IF first-order logic.

Thus the prima facie evidence for or against the applicability of my type of truth definition to natural languages is the reverse of what it is for Tarski-type truth definitions. It suggests strongly a positive answer for my definition and a negative one for Tarski's definition.

One can say more than this, however. In a sense, my truth definition for IF first-order languages provides a paradigmatic example of a truth definition for a language with a relatively poor syntax, which through its very form shows how it can be extended to syntactically richer languages or language fragments. In order to see this, let us have a look at the truth predicate itself in its second-order form. Its initial clauses tell essentially that for atomic sentences we are adopting a redundancy (disquotational) treatment of truth. The other clauses specify the conditions of the existence of a winning strategy for the initial verifier for sentences containing quantifiers and propositional connectives. Hence the upshot of the truth predicate seems to be merely to spell out how the truth of an IF first-order sentence depends on the quantifiers and propositional connectives it contains. Hence it might seem that my truth predicate can be defined only for quantificational languages, IF or not, and it apparently has little relevance to other kinds of languages. The treatment of quantifiers can be extended to natural languages without any problems in principle, even though the actual logical behavior of ordinary language quantifiers is quite different from that of the formalized quantifiers of first-order languages. But at first sight it is hard to see how my treatment could be extended to richer languages.

This objection is inconclusive, however. Assume, for the sake of the argument, that we enrich a first-order language further by introducing fresh logical constants. Assume, likewise for the sake of the argument, that these new constants admit of a game-theoretical treatment along the same lines as quantifiers and propositional connectives; that is to say, by means of an affective game rule which syntactically speaking simplifies the sentences that the players are considering. Assuming that an application of a game rule by the verifier can be thought of as an attempted step toward verification, then the definition simply spells out the contribution of different logical and nonlogical constants to the truth-conditions of the language in question. Hence it should not be any surprise, and much less an objection, that for languages which have been developed for the explicit purpose of studying the semantics of quantifiers and connectives, the first-order languages, only the contribution of quantifiers and propositional connectives is spelled out in so many words. What my truth definition provides is therefore a paradigm case of truth definitions, an example whose purview is automatically extended when other logical and nonlogical constants are brought within the scope of semantical analysis. The crucial idea of truth as the existence of a winning strategy remains completely unmodified. What such extensions require is the possibility of extending the game-theoretical treatment to logical and nonlogical constants other than quantifiers and propositional connectives. There are enough examples offered by game-theoretical analyses of different natural-language expressions to suggest strongly that such a treatment is possible over a wide front of semantical phenomena. Such analyses as can already be found in literature include anaphora, genitives, prepositional phrases, questions and answers, epistemic verbs, and so forth (cf. here Hintikka and Kulas 1983 and 1985).

There is another, independent line of argument for the general significance of the truth definitions explained in this chapter. A partial answer can be formulated in terms of the power and central role of IF first-order languages. In twentieth-century philosophy and logic, ordinary first-order logic has often been taken to be *the* basic logic. Sometimes it has even been claimed that it is the logic of ordinary language. Frege thought that his *Begriffsschrift* is a universal "conceptual notation" or *mathesis universalis* in Leibniz's sense. Frege, and most of his followers, have presented first-order languages as resulting from a minor regimentation of ordinary language, involving mostly the elimination of the unfortunate ambiguities and other imperfections which beset the unpurified language of our tribe. Admittedly, Frege's logic is not a pure first-order logic, but it contains higher-order ingredients. But since he can be shown to have assumed a nonstandard interpretation of the higher-order part of his logic, he might as well have construed his *mathesis universalis* as a many-sorted first-order language (see Hintikka and Sandu 1991). Moreover, it will in any case be seen in Chapter 9 that one can in effect do everything in the IF first-order logic that can be done in a higher-order logic.

Among other things, it is in a sense assumed widely these days that all of mathematics can be done by means of ordinary first-order logic for it is widely believed that all mathematics can be done in terms of an axiomatic set theory. Moreover, this axiomatization is typically taken to be a first-order one. This means that the only logic it involves is the usual first-order logic. The other elements, including the axioms, are according to this view specifically mathematical rather than logical, thus leaving first-order logic as the only logic of mathematics.

Moreover, Frege's (and his followers') claim that ordinary first-order logic is our true *Sprachlogik* seems to be supported by Chomskian linguistics. Chomsky's recent version of the old idea of logical form, teasingly called just LF, is essentially like a formula of first-order logic (see here, e.g., Chomsky 1986 and cf. Hintikka 1989). The few steps beyond ordinary first-order logic that Chomsky or his followers have taken go unsurprisingly in the direction of IF first-order logic.

Even if these arguments have an aura of *ad hominem* around them, others are readily forthcoming. In Chapter 9 it will be seen that in a perfectly good sense most of ordinary mathematics can be done in IF first-order languages. I have shown earlier that informationally independent quantifiers occur in natural languages. In the next chapter, it will be seen that negation behaves in natural languages very much like the negation (or negations) of IF first-order language.

Even though such arguments are suggestive rather than binding, they do show at the very least the very general interest of my truth predicate.

Philosophically the most important upshot of the results of this chapter is undoubtedly that they disprove, once and for all, the myth that the notion of truth in a sufficiently strong language is inexpressible in that language itself. Applied to our actual working language, this myth entails the companion myth of the ineffability of truth in the ordinary sense of the word. These myths are ready to be discarded. Not only has it been shown in this work that such a self-applicable notion of truth is possible in some recondite formal languages – the construction presented above can serve as a paradigm case for the definition of a truth predicate for any language whose logic contains IF first-order logic. And this logic is, as I have argued, nothing but our most basic ground-floor elementary logic. Hence the strong suggestion of what has been seen here is virtually the contrary of the ineffability myth. As soon as you understand your own language and its logic, you have all that it takes to understand and even to define the concept of truth, or so the suggestion goes.

This general point can be illustrated with reference to a more specific philosophical problem. It may be called the two hats problem. These two hats are apparently worn at the same time by my semantical games for first-order logic. The language games of seeking and finding are in the first place games for quantifiers and propositional connectives. The player's moves in them are governed by the rules that characterize the meaning of quantifiers and connectives. Semantical games of the kind I have described are the logical home of quantifiers and connectives.

But if so, how can these very same games also serve to give an altogether different kind of concept its meaning, namely, the concept of truth – at least the notion of truth as applied to first-order languages. How can one and the same language game serve to lend a meaning to two different kinds of concepts, one of which (the concept of truth) seems to be a metalogical one? This two hats problem might also be called Wittgenstein's problem (cf. here Hintikka and Hintikka 1986, Ch. 1). For Wittgenstein insisted that you cannot speak meaningfully and nontrivially of the truth of the sentences of a language in that language itself. Or, since for Wittgenstein there is ultimately only one language ("the only language that I understand"), we cannot speak of truth nontrivially, period. What looks like a metalogical discourse pertaining to the truth and falsity of a fragment of language is for Wittgenstein merely a different "calculus," a different language based on a different language game. How, then, can the meaning of first-order logical constant *and* the notion of truth as applied to first-order languages be constituted by the same language games? Doesn't speaking of truth take us ipso facto to a metatheoretical level?

A clue to an answer to this question is the close connection there is between the notions of truth and meaning. What a sentence says is that its truth-conditions are satisfied. Hence it is impossible to try to detach the notion of truth as applied to a quantificational sentence to the language games that lend quantifiers their meaning.

This answer can be elaborated further. For the purpose, it is handy to recall the distinction between *definitory* rules of a game from the *strategic* rules or principles governing it. The former specify which moves are permitted and which ones are not. The latter, the strategic rules, specify which strategies (and a fortiori which rules) are better or worse than others. The several definitory rules of semantical games characterize the meaning of the different quantifiers and propositional connectives. In contrast, we have seen that the notion of truth for quantificational sentences is to be characterized in terms of the sets of strategies that are open to the initial verifier to use. Thus the step from the meanings of quantifiers to the truth of quantificational sentences is not a step to a metalogical level. It is, rather, a step from considerations pertaining to the particular moves in a semantical game to concepts pertaining to the strategies of the initial verifier in those games.

The reason why it is impossible to detach the meaning of quantifiers from the notion of truth is that some degree of the mastery of the strategies of a game is unavoidable for the understanding of a game. If you only know how chessmen are moved on the board you cannot as much as say that you know how to play chess, unless you have some idea of the better and the worse moves and sequences of moves. No one would deign to play chess with you. This illustrates the fact that some degree of understanding of the strategies of a game is an integral part of the conceptual mastery of that game.

But if so, we then have here an example of the general insight mentioned above. Understanding a quantificational language means mastering the language games that give quantifiers their meaning. Now, such mastery was just seen to involve a grasp of the strategies available to the players of a semantical game. But those strategies are just what is needed to understand the concept of truth as applied to quantificational languages.

One of the main features of the truth predicate presented here is that it is in *first-order* terms. As was pointed out previously, Tarski-type truth definitions are of a Σ_1^1 *second-order* form. The reason why this makes a crucial difference to the foundations of logic and mathematics was explained in Chapter 1. It was explained there in what sense all model theory depends on truth definitions. As long as these definitions can only be given on second-order level, then model theory depends on second-order logic. And even if second-order logic is not merely "set theory in sheep's clothing", it still admittedly involves many of the same problems of set existence that makes set theory not only a difficult but positively frustrating exercise.

Against this backdrop, one can see the significance of my truth predicate. It is nothing more and nothing less than the declaration of independence of model theory. It shows that *one can develop a model theory for the powerful IF first-order languages on the first-order level*, ergo independently of all questions of sets and set existence. All the quantifiers in the IF first-order version of my truth predicate range over individuals. If Quine is right (as I think he is, if rightly understood) and to be is to be a value of a quantified variable, then our truth definition (and indeed the entire IF first-order logic) does not take up the question of the existence of higher-order entities like sets at all. Consequently, all apprehensions concerning the purely logical status of model theory are groundless. Model theory of first-order logic is part of logic, and not a proper part of mathematics. The problems which are caused by the apparent dependence of the model theory of first-order logic on set theory (or on higher-order logic) can thus be solved, and Tarski's curse be exorcised. This makes a substantial difference to the foundations of mathematics. If you return to Chapter 5 and review what was found there, you can see in how many different ways freedom from Tarski's curse enhances the self-sufficiency of first-order logic as contrasted to set theory.

Of course, my saying this does not mean that I advocate doing model theory in everyday practice in IF first-order languages. That would be impracticable. What is at issue is the conceptual nature of model theory and not the heuristic framework best suited for the actual nitty-gritty work in model theory.

The conceptual independence of model theory from set theory is only one aspect of a more comprehensive reevaluation that is needed in our ideas about the foundations of logic and mathematics. This reevaluation will be continued in Chapter 8.

The philosophical consequences and suggestions of the definability of a truth predicate in suitable IF first-order languages are too sweeping to be exhausted here. Let me mention only a couple of new tacit perspectives. Suppose that it could be shown that our ordinary language (or some significant fragment thereof) is inevitably like IF first-order languages as far as the notion of truth is concerned. This is suggested by what has been found in this chapter, and it will be reinforced by what is found in Chapter 7. Then the definability of truth would be equally inevitable, and hence it could be argued that merely by using our own language we are inextricably committed to a realistic concept of truth, namely, the one definable along the lines I have followed. This would be an important conclusion indeed. Even though it is beside the main purpose of this book to examine such issues here at any greater length, I can point out that this line of thought promises to support very strongly what was said at the end of Chapter 2 of the independence of our normal notion of truth from the epistemic institutions of verifying and falsifying propositions. Truth is in a deep sense a logical rather than an epistemological notion.

To put the same point in other terms, anyone who is using a language with a minimally rich expressive power (logically speaking) is committed to a concept of truth which in the current crude classification would be called a correspondence view of truth. This would be the case even if that language were only used for the purpose of Rortian dialogues for dialogue's sake. Any sufficiently strong language would be a mirror of the world, Rorty notwithstanding.

Notes

1 The only book-length discussion of Tarski's truth-definition seems to be Moreno (1992). It supplies further references to the literature.

2 See Walkoe (1970).

3 For infinitely deep languages, see the bibliography in the introduction to Tuuri (1990). For game quantifier languages, see Ph. G. Kolaitis, "Game Quantification", Chapter 10 in Barwise and Feferman (1985).

7

The Liar Belied: Negation in IF Logic

(A) Tertium datur

So far so good. A truth predicate can after all be defined in an IF first-order language for the Gödel numbers of the sentences of that very same language. But what about the arguments which Tarski and others have presented against the definability of truth for a formal language in that language itself? The best known of these arguments are closely related to the time-honored or perhaps rather time-abused liar paradox (for this paradox, see Martin 1978 and 1984). They are illustrated by an outline of an argument which establishes the indefinability of truth for a first-order arithmetical language in that language. Such an argument can use the technique of Gödel numbering. In rough outline, the argument runs as follows:

Let **n** be the numeral representing the number n. According to the diagonal lemma, for any expression $S[x]$ in the language of elementary arithmetic containing x as its only variable, there is a number n represented by the numeral **n** satisfying the following condition:

$$g(S[\mathbf{n}]) = n \tag{7.1}$$

where $g(S[\mathbf{n}])$ is the Gödel number of the expression obtained from $S[x]$ through replacing x by **n**. Assume now that there is a truth predicate $T[x]$ in the arithmetical language in question, that is, a numerical predicate $T[x]$ such that $T[\mathbf{n}]$ is a true arithmetical statement if and only if n is the Gödel number of a true arithmetical statement. Then we could apply the diagonal lemma to $\sim T[x]$ and obtain a number d such that the Gödel number of

$$\sim T[\mathbf{d}] \tag{7.2}$$

131

is d. Hence what (7.2) says is that the sentence with the Gödel number d is false. But that sentence is (7.2), which yields a contradiction.

This line of thought is similar to the usual proofs of Gödel's incompleteness theorem except that there the actual provability predicate and not the hypothetical truth predicate is being considered.

What happens to this highly important line of thought when we move from ordinary first-order arithmetic to IF first-order arithmetic? At first sight this argument might seem impossible to default. After all, IF first-order logic is stronger than the ordinary one, and hence the paradox might seem to be impossible to avoid also in IF first-order languages.

In order to see why the liar paradox does not arise in IF first-order languages, we have to go back to the basic ideas of game-theoretical semantics. In it, the truth of a sentence S is defined as the existence of a winning strategy for the initial verifier in the corresponding semantical game $G(S)$. The falsity of S is defined as the existence of a winning strategy of the initial falsifier in $G(S)$. Negation is handled by stipulating that $G(\sim S)$ is like $G(S)$ except that the roles of the two players have been exchanged.

These stipulations are simple and natural, but they have a striking consequence. This consequence concerns the law of excluded middle or the law of bivalence, as some people prefer to put it. This principle of *tertium non datur* becomes a determinacy assumption in the sense of game theory. In other words, it asserts that there always exists a winning strategy for one or the other of the two players of a semantical game. As we know from game theory (and from set theory), determinacy assumptions are normally far from obvious. They often fail. They represent strong assumptions which allow a quantifier inversion, in that a determinacy assumption enables one to infer from the nonexistence of a winning strategy for one of the players the existence of a winning strategy for the other one. Sight unseen, there is therefore no reason to think that the law of excluded middle should hold in my semantical games. It is only by a lucky coincidence, so to speak, that it holds in ordinary first-order logic. In IF first-order logic, it is easily seen to fail, and to fail in a radical manner.

In order to see this, we can recall that many of the "nice" metatheorems that hold in ordinary first-order logic continue to hold in IF, sometimes in a strengthened form. One of them is the separation theorem, which says that two jointly inconsistent but separately consistent sets of formulas of an IF first-order language, say σ and τ, can always be "separated" by a *single ordinary first-order formula F* in the sense that $\sigma \vdash F$ and $\tau \vdash\, \sim F$.

Assume now that the contradictory negation of an IF first-order sentence S_2 is somehow expressible in the same language, say by the sentence S_1. Then by applying the separation theorem to $\{S_1\}$ and $\{S_2\}$ it is seen that both of them must be sentences of ordinary first-order logic. For then there is a separation formula S_0 which belongs to an ordinary first-order language such that

$$S_1 \vdash S_0$$

$$S_2 \vdash\, \sim S_0$$

But since $\vdash (S_2 \leftrightarrow\, \sim S_1)$ it follows that S_1 is equivalent to S_0 and S_2 to $\sim S_0$. This shows that the only IF first-order formulas whose contradictory negation is expressible in the IF language are precisely the ordinary first-order formulas. In this sense, the law of excluded middle holds only in the fragment of an IF language consisting of ordinary first-order language.

The upshot is that in IF first-order languages, the law of excluded middle inevitably fails. This result requires a number of comments and explanations.

(i) This result is not due to our arbitrary choice to define falsity in a certain way for IF first-order languages, namely, in the way explained above. It is independent of our terminology. The real issue is how negation behaves in IF languages. Indeed, whatever interest my observations may have is due to the fact that we did not assume any breakdown of the law of excluded middle or any truth-value gaps or a third truth-value. What we have is an inevitable consequence of the most natural semantics for IF first-order languages, namely, the game-theoretical one. Indeed, since neither Tarski-type truth definitions nor substitutional ones are available in IF first-order languages, we scarcely have any choice here but to use some version of game-theoretical semantics.

(ii) This observation can be sharpened. Let us merely assume that the usual interconnections hold between negation and the other logical constants. These interconnections are codified in the law of double negation, in De Morgan's laws and in the interdefinability of the two quantifiers. These laws certainly hold for classical negation.

Assume also that the usual game-theoretical rules for the other notions (&, $\vee$, $\exists$, $\forall$) apply. They, too, merely incorporate perfectly classical assumptions. We then need a rule for negation only for the trivial case of a negated atomic sentence. Yet the result is the same treatment for negation as was outlined above. Inter alia, the law of excluded middle fails. Somehow an apparently nonclassical logic results from entirely classical assumptions.

Thus we can see that the prima facie surprising behavior of negation in IF first-order logic is not due to our game rule for negation, which reverses the roles of the two players. This rule can be replaced by the perfectly classical rules which merely push negation signs deeper into a formula. Given the other game rules, the two kinds of rules for negation are simply equivalent.

In a sense illustrated by this line of thought, we cannot help treating negation in the way indicated above, which leads to the inevitable result that $\sim S$ is not the contradicatory negation of S but a stronger (dual) negation.

(iii) The kind of failure of the principle of excluded middle that we are dealing with here is an unavoidable combinatorial consequence of the way quantifiers and other concepts interact with each other. It has nothing to do with the limitations of human knowledge. Epistemic failures of *tertium non datur* should be studied in epistemic logic, not here. The failure I have described has nothing to do with some particular subject matter, for example, with the future as distinguished from the past. It has nothing to do with infinity, either. Below (see (v)) a finite miniexample is given of a sentence pertaining to a universe of six individuals where one can literally see (at least after a few moments thought) that a certain sentence is neither true nor false. You can come to see virtually directly that any attempt to verify a certain simple sentence can be defeated and that any attempt to falsify it can likewise be defeated. What else can any man (or, as Samuel Johnson might add, any woman or any child) reasonably say here except that such a sentence is neither true nor false? Thus a little reflection can easily bring home to you the naturalness, nay, the inevitability of our treatment of negation.

(iv) Unlike other approaches, such as intuitionism, our treatment of a logic in which the law of excluded middle fails does not involve any tampering with the definitory laws of ordinary first-order logic. The game rules for semantic games connected with IF first-order sentences are precisely the same as those for ordinary first-order

languages, except for allowing that certain moves are made by the
initial verifier in ignorance of certain earlier ones. Moreover, the
logical truths and valid rules of inference of ordinary first-order logic
remain unproblematically valid.

(v) It might be instructive to illustrate these remarks by a simple
example. For the purpose, consider a miniuniverse consisting of six
entities, three gentlemen (Alan, Brian, and Cecil) who between them
have three hobbies, namely, riding, sailing, and tennis. Indicating
hobbies by an arrow, we can sum up the relationships holding here
by means of the following diagram:

$$
\begin{array}{l}
\text{Alan} = a \bullet \longrightarrow \bullet\, r = \text{riding} \\
\text{Brian} = b \bullet \times \bullet\, s = \text{sailing} \\
\text{Cecil} = c \bullet \longrightarrow \bullet\, t = \text{tennis}
\end{array}
\qquad\text{(D)}
$$

For simplicity, I will tacitly restrict the variables x and z to gentlemen and the variables y and u to hobbies. Consider, then, the
following statements:

$$(\forall x)(\forall z)(\exists y)(\exists u)(H(x, y) \,\&\, H(z, u) \,\&\, (y \neq u)) \tag{7.3}$$

$$(\forall x)(\forall z)(\exists y/\forall z)(\exists u/\forall x)(H(x, y) \,\&\, H(z, u) \,\&\, (y \neq u)) \tag{7.4}$$

$$(\exists x)(\exists z)(\forall y)(\forall u)((H(x, y) \,\&\, H(z, u)) \supset (y = u)) \tag{7.5}$$

Here (7.3) says that any two different gentlemen have respective
hobbies that are different, that is, that no two different gentlemen
have all their hobbies in common. Moreover, by putting $x = z$ we see
that (7.3) implies that each gentleman has at least two hobbies. It is
immediately seen that (7.3) is true in the model (D). It is also seen that
(7.5) is the contradictory negation of (7.3) and hence false in (D).

Clearly, (7.4) is true if and only if there are function f and g such
that

$$(\forall x)(\forall z)(H(x, f(x)) \,\&\, H(z, g(z)) \,\&\, (f(x) \neq g(z))) \tag{7.6}$$

Can such functions be defined in the model D? Let us see. By
symmetry it suffices to consider the case

$$f(a) = r \tag{7.7}$$

Then a substitution of a for both x and z in (7.6) yields

$$H(a, f(a)) \,\&\, H(a, g(a)) \,\&\, (f(a) \neq g(a)) \tag{7.8}$$

This is possible only if

$$g(a) = s \tag{7.9}$$

By putting $x = a$, $z = b$ we obtain

$$H(a, f(a)) \mathbin{\&} H(b, g(b)) \mathbin{\&} (f(a) \neq g(b)) \tag{7.10}$$

This is possible only if

$$g(b) = t \tag{7.11}$$

In the same way as in (7.9), it can be shown that

$$f(b) = r \tag{7.12}$$

In the same way as in (7.11) we can show that

$$f(c) = s \tag{7.13}$$

But then we can substitute c and a for x and z, respectively, and obtain

$$H(c, f(c)) \mathbin{\&} H(a, g(a)) \mathbin{\&} (f(c) \neq g(a)) \tag{7.14}$$

which is in view of (7.7) possible only if

$$f(c) \neq s \tag{7.15}$$

But this contradicts (7.13), showing that (7.4) is not true in D.

By reviewing this line of thought a couple of times you can reach a point at which you can directly see the failure of (7.4) to be true simply by inspecting the model. In other words, you can train yourself literally to see from the model (D) that any strategy of attempted verification of (7.4) can be defeated by a suitable counterstrategy.

Now, it can easily be seen that (7.5) is not only the (dual) negation of (7.3) but also of (7.4). The reason is that a winning strategy must work no matter what moves my opponent makes, independently of their dependence on (or independence of) others. Hence neither (7.4) nor its negation (7.5) is true – that is, the law of excluded middle fails for (7.4).

This simple example illustrates several important facts, some of them explained above. It shows that the failure of the law of excluded middle is by no means assumed in our treatment. It is a consequence of a certain combinatorial feature of the behavior of negation in IF logic. In intuitive terms, this combinatorial fact can be interpreted as saying that every serious attempt to verify (7.2) in (D) can be defeated and that every serious attempt to falsify it in (D) can be likewise defeated. By serious attempt, I mean here an attempt governed by an explicit strategy. In the case on hand, the relevant combinatorial fact is a finitary one. This shows that infinity cannot in any way be blamed for the failure of *tertium non datur*, even if in other cases the combinatorics involved is infinitary.

To put the same point in a different way, the applicability or nonapplicability of *tertium non datur* to a given sentence in a certain model ("world") is an objective fact about that model or world. It does not depend on any limitations of one's knowledge about the world, nor does it imply that there are such limitations.

More generally, the example illustrated the fact that questions concerning the logical truth and satisfiability of IF first-order formulas are in a reasonable sense *combinatorial*. They are questions concerning the possibility of certain kinds of structures of individuals. These "kinds" are characterized in a way similar to, say, the structures involved in Ramsey-type theorems. These structures can be finite, but they may of course also be infinite. In neither case are any totalities of sets or functions or any other kinds of higher-order entities involved.

What distinguishes our example from others is its simplicity. By looking at (D) and by reviewing the argument presented above in your mind, you can reach a point where you can almost literally see the failure of the principle of excluded middle. Even though the same cannot be said of more complex cases, the difference lies merely in a contingent limitation of human capacities, and not in the subject matter.

It is even possible in principle to formulate examples of sufficiently simple infinite structures where the failure of *tertium non datur* could be seen equally directly.

All this illustrates the fact that our truth definition deals with the strategies of semantic games and not with their definitory rules. Later, in Chapter 10, an important possibility will be found of changing our truth definition without changing the move-by-move definitory rules of semantical games.

Here the reader is perhaps already catching a glimpse of why it is hopeless to try to implement constructivistic ideas by somehow manipulating the definitory rules of the games of formal proof. Even when we move from such formal games to their real basis in the form of semantical games it still does not help to try to vary the definitory rules of such games. Conversely, our results illustrate at the same time the advantages of a strategic viewpoint in logical and linguistic theory.

(vi) In order to avoid misunderstanding, it is in order to point out that the failure of the law of excluded middle for a certain sentence S is no guarantee that its contradictory negation is not expressible in a IF first-order language. For instance, it is easily seen that in many models a sentence of the following form is neither true nor false:

$$(\forall x)(\exists y/\forall x)A[x, y] \qquad (7.16)$$

where $A[x, y]$ is atomic. Yet (7.16) is easily seen to be logically equivalent to

$$(\exists y)(\forall x)A[x, y] \qquad (7.17)$$

and hence to have as its contradictory negation

$$(\forall y)(\exists x) \sim A[x, y] \qquad (7.18)$$

What happens in the more complicated example just presented is not only that *tertium non datur* fails, but that the contradictory negation of a sentence is not expressible in the IF first-order language in question.

The behavior of negation in IF first-order languages has a number of consequences, including the following:

(a) Unless conditionals are treated separately, material implication (truth-functional conditional) is not expressible in general in IF first-order languages. For instance, $(F \supset G)$ and $(\sim F \vee G)$ express something stronger than the customary material implication of truth-function theory.

   It would be interesting to examine whether this fact might have tacitly been responsible for the unease many philosophers have felt about truth-functional conditionals.

(b) The same applies to equivalences. If $(S_1 \leftrightarrow S_2)$ is construed as

$$(S_1 \ \& \ S_2) \vee (\sim S_1 \ \& \ \sim S_2) \qquad (7.19)$$

then its logical truth entails that S_1 and S_2 both satisfy the law of excluded middle and hence belong to ordinary first-order logic.

(c) As an application of these observations, consider Tarski's T-sentences. They are substitution-instances of the schema

   (T) Π is true $\leftrightarrow p$

where "Π" is a placeholder for a quote or a structural description of the sentence which is to replace "p." (T) is equivalent to

(T)* (Π is true & p) ∨ (Π is not true & $\sim p$)

(T)* clearly entails

$$p \vee \sim p \qquad\qquad (7.20)$$

In other words, any sentence whose truth can be characterized by means of an explicitly formulated instance of the T-schema satisfies the law of excluded middle; it has a contradictory negation in the same language. But if so, Tarski's T-schema is useless in its usual form in IF logic. What it tries to express is true and important, namely, that a sentence S and its truth-condition $c(S)$ must be true simultaneously. But the usual formal expressions of the T-schema require more. It requires that S and $c(S)$ must also be false simultaneously. This requirement cannot be sustained in IF first-order logic. Hence the entire matter of the T-schema has to be handled with caution. It is unproblematic only in ordinary first-order languages.

As I have pointed out before (Hintikka 1976b), the T-schema fails for another reason, too. Like other features of Tarski's approach, it presupposes compositionality. When compositionality fails, T-schema may also fail. Cases in point are found when the T-schema is formulated in natural languages, for example, as follows:

(T)** Π is true if and only if p

with "Π" and "p" working as before, but this time "p" working as a placeholder for English sentences (free of indexicals). By substituting "anyone can become a millionaire" for "p" and its quote for "Π" and by taking only one half of the equivalence we can see that (T)** entails

"anyone can become a millionaire" is true if anyone can
    become a millionaire   (7.21)

But (7.21) is not true: one person making a cool million does not imply that everyone can do it.

The explanation is that the game rule for *any* has priority over the game rule for conditionals. From that it follows that the force of *any* depends on where it occurs in a sentence, for instance, in a conditional sentence. In other words, it violates compositionality.

Because of this scope peculiarity of *any*, it has a wider scope in (7.21) than *if*, thereby making (7.21) false.

This counterexample to Tarski's T-schema might seem to turn on a peculiarity of the English *any* that has no general theoretical interest. That would be a misperception, however. What (7.21) really illustrates is the failure of compositionality in natural languages, ergo an extremely interesting general phenomenon.

In sum, the T-schema is virtually useless for a discussion of truth and truth definitions in general. The presuppositions it is predicated on cannot be expected to hold in general outside ordinary first-order languages.

The failure of the T-schema to serve as a satisfactory criterion of the characterization of truth also means that disquotational accounts of truth are seriously incomplete. They presuppose compositionality and *tertium non datur*, neither of which can be expected to hold in languages of more than minimal expressible power and surely not in ordinary language. Consequently, Tarski's T-schema cannot serve as a condition on our conception of truth in general.

The failure of truth-functional conditions to be expressible in IF first-order languages is a more significant fact than one might first suspect. This significance can be seen by asking: What do we need truth-functional conditionals for, anyway? An important part of the answer is: We need them for the purposes of logical proof. For instance, without a contradictory negation at our disposal, we cannot use *modus ponens*, except in a weakened form.

The same point can be put in another way. The formulas that can serve to justify inferential steps from one sentence to another, as the truth-functional conditional "if S_1, then S_2" would serve to facilitate an inference from S_1 to S_2, turn out to be extremely strong sentences semantically. They cannot be expressed in ordinary first-order language, and they cannot be expressed in IF first-order language, either. Later in this chapter we shall see an example of a still stronger language in which they can be so expressed.

This strength and subtlety of inferential relationships, reflected by their failure to be expressed except in very strong languages, makes it highly unlikely that we humans could have direct unaided insights into their status as "inference tickets". And if so, another common view of logic bites the dust. It is the idea of logic as a systematization of our intuitions about relationships of logical consequence between propositions. There is precious little reason to think that we could have viable "intuitions" about matters so complex that they cannot even be expressed in an IF first-order language.

This observation further enhances the status of model-theoretical concepts and theories as compared with the role of proof-theoretical (inferential) approaches to logic.

An additional observation is relevant here. In Chapter 3 it was seen that IF first-order logic enables us to capture, as the models of a formula, classes of structures that cannot be likewise captured by any ordinary first-order formula. Let Σ be such a class of structures, captured by the formula $F(X)$. Then it follows from what has been seen that the complement of Σ cannot be captured by any IF first-order formula. For this complement would have to be captured by the contradictory negation of $F(X)$, which is not available unless $F(X)$ is a formula of ordinary first-order logic. But then it does not enhance the representational capacities of ordinary first-order logic.

The behavior of negation in IF first-order logic also shows how this logic can escape the clutches of the third main result (besides Gödel's and Tarski's) that apparently severely restricts what can be done by means of logic. This result is known as Lindström's Theorem (see Lindström 1969; cf. Ebbinghaus, Flum and Thomas 1984, Ch. 12). It is a result in the theory of abstract (model-theoretical) logics which says, roughly speaking, that the ordinary first-order logic is in a natural sense the strongest possible logic satisfying certain conditions. This sense is simply an explication of the idea that a stronger logic enables one to make finer distinctions among models than a weaker one.

Of the conditions of Lindström's Theorem, most attention has been directed at the assumptions that the logic in question is compact and that it validates the Löwenheim–Skolem theorem. Since IF first-order logic satisfies these conditions and yet is stronger than it in the relevant sense of the term, these two assumptions cannot be all the important ones. Indeed, the joker in the pack turns out to be Lindström's assumption that negation behaves like contradictory negation. This assumption is not satisfied by IF first-order logic, and for this reason it can be stronger than ordinary first-order logic, Lindström's Theorem notwithstanding. Thus negation once again plays a crucial role in the foundations of logic. And, once again, apparent restrictions on the power of logic turn out not to be real ones.

(B) The liar is merely a nontruth-teller

In the light of our results, we can now see why it is that the liar paradox does not arise in IF first-order languages. The status of the diagonal lemma strictly speaking needs a review. But since it basically expresses a combinatorial fact about elementary number theory, there is no reason to expect that it does not hold in IF first-order logic. But what escape hatches does its validity leave us with?

Let us see. Let us apply the diagonal lemma to our truth predicate T. Then we obtain the result that there exists a number h such that h is the Gödel number of

$$g(\sim T[\mathbf{h}]) \tag{7.22}$$

Then it is easily seen in the usual way that if

$$\sim T[\mathbf{h}] \tag{7.23}$$

is false, it is true, and vice versa. But (7.23) need not be either true or false. Hence no contradiction ensues. The liar paradox does not vitiate the truth definition we have given for suitable IF first-order languages. The liar sentence merely turns out to be neither true nor false.

This resolution of the liar paradox is so simple as to prompt surprise and even disbelief. A few comments are therefore in order.

(a) There is nothing mysterious about the fact that (7.23) is neither true nor false. That it is neither is merely a combinatorial feature of the structures that (7.23) speaks of. In principle, this fact is as straightforward as our example about (D) above. It means that each strategy of verification for (7.23) can be defeated and that each strategy of falsification can likewise be frustrated. That we cannot see this fact directly, as we could see the corresponding fact in the case of (7.4) in relation to (D), is merely due to the greater complexity of (7.23) as compared to (7.4). Otherwise the two are on a par. One might even suggest that the dramatization of the properties (7.23) via a comparison with the liar paradox is a partial way of highlighting this intuitive feature of (7.23). My point here is closely related to the fact that Gödel's impossibility theorem, which he himself compared with the liar paradox, can be proved constructivistically.

(b) Furthermore, the combinatorial features of (7.23) are independent of our terminology. They cannot be changed, for example, by defining falsity in a way different from ours. The crucial point is that there is basically only one way of characterizing negation in IF first-order logic.

It has sometimes been suggested (cf. e.g., Simmons 1990, p. 290) as a solution for paradoxes like that of the liar that there is in the last semantical analysis no paradox-generating concept to be expressed. What has been seen shows that this is not the case in the approach developed here. For the existence of a winning strategy for the initial verifier is a definite combinatorial property of models, no matter whether the game in question is determinate or not. Hence, in a very concrete sense, there is such a concept of truth to be defined for IF first-order languages.

(c) It was assumed above in the argument concerning (7.22)–(7.23) that the diagonal lemma applies when elementary arithmetic is developed within the framework of IF first-order logic instead of an ordinary one. Even though the correctness of this assumption is fairly obvious, it is still worth noting here. The reason why the diagonal lemma applies is that the syntax of IF first-order arithmetical language can be handled (via a suitable Gödel numbering) in its own ordinary first-order-logic fragment. Hence one can represent the diagonal functions (cf. Mendelson 1987, p. 155) in the language as before, for it deals simply with the syntax of the arithmetical language in question. Furthermore, the representability of the diagonal functions is the only substantial assumption needed for the diagonal lemma, which is therefore applicable in IF first-order arithmetical languages.

The same remarks apply *mutatis mutandis* to the Σ_1^1 fragments of second-order arithmetical languages, unsurprisingly, because they are equivalent to the corresponding IF first-order languages.

(d) If you take a wider view of the situation, you will find that not only should you have expected the liar paradox not to touch our truth definition but that you should have expected it to be immune to all paradoxes. In order to see this, you may have a look at the second-order truth definition as presented in Chapter 6. What it says is that there exists a predicate (of the Gödel numbers of sentences) satisfying certain conditions such that x has this predicate. What are the relevant conditions? They spell out precisely the features which our intuitive concept of truth has, applied to the given IF first-order language *sans* any special truth predicate. There is no self-reference here to generate paradoxes or contradictions. Unless our intuitive conception of truth is seriously flawed, this second-order definition cannot go astray, for it

144 THE PRINCIPLES OF MATHEMATICS REVISITED

merely captures what we mean by the truth of an IF first-order sentence. Hence no paradoxes can be generated by such a definition. And the possibility of translating the second-order truth definition back into the IF first-order language in question is simply a matter of a well-known logical result.

(e) The ease at which the liar paradox was avoided in IF first-order languages suggests that it does not pose a serious threat against truth definitions or against the possibility of a consistent concept of truth also in natural languages. Instead, it suggests an important general point. In some sense, the liar paradox has more to do with our concept of negation than with our concept of truth. This is suggested also by the way in which liar-like arguments are used by Gödel and Tarski. At bottom, these uses are based on the diagonal lemma which is a combinatorial fact about first-order arithmetical languages. There is no leeway with respect to such a metatheorem. A fortiori, any truth predicate is vulnerable to the diagonal lemma. There is no way of avoiding this application. Hence the only possible way out is to challenge the true-false dichotomy.

In a constructive vein, the failure of the liar paradox in IF first-order languages, notwithstanding the possibility of being able to define their own truth, throws some further light on the prospects of giving truth definitions for natural languages. Indeed, liar-like paradoxes and the diagonal arguments which they rely on are often adduced as a reason for the impossibility of defining truth for natural languages.

What has been presented is more than just an example of a rich language for which truth can be defined in the language itself. Independently of all questions concerning the relations of IF first-order languages and all questions as to whether any truth definitions like the ones expounded here are possible in natural languages, the definability of truth for IF first-order languages has certain massive consequences. It explodes once and for all the myth that semantical concepts like truth cannot be expressed in the (sufficiently rich) language to which they are applied, as well as the myth that a Tarski-type hierarchy of metalanguages is inevitable for the purpose of escaping semantical paradoxes, as has been alleged (cf. Simmons 1990, especially pp. 296–299).

Our results incidentally show that the main competing approaches to truth fail to do the same job as the account offered here. This holds both of redundancy "theories" of truth including disquotational ones, and of inductive truth definitions. If they are formulated in the same language as the one to which the proposed truth predicate pertains, then they are subject to the liar-type counterexamples discussed here.

(f) A wary reader might nevertheless still feel insecure about the dangers of possible paradoxes. One source of worry is that, even though IF first-order logic is (semantically) incomplete there still exists a complete disproof procedure. Does this completeness perhaps make IF first-order logic too strong to be safe?

The answer is that the complete disproof method does not give rise to any paradoxes, but instead gives rise to interesting results. For the purpose, consider a nonlogical finitely axiomatizable theory T which contains elementary arithmetic and which is formulated in an IF first-order language. The general disproof procedure for IF first-order logic yields a disproof procedure for the theory T, in the sense of a complete proof procedure for establishing inconsistency with the axioms of T. Using this technique of Gödel numbering, we can discuss the metatheory of T in T itself. The complete disproof procedure just mentioned gives rise to a predicate $Disp(x)$, which is true of the Gödel number $g(S)$ of a sentence S iff S is disprovable on the basis of T, assuming that T is consistent. Because of the completeness of the procedure, $Disp(x)$ holds of the Gödel number $g(S)$ of S if and only if S is inconsistent with T. If we apply the diagonal lemma to the formula $Disp(x)$, we obtain a formula

$$Disp(\mathbf{n}) \tag{7.24}$$

where the Gödel number of (7.24) is n and where $\mathbf{n}$ is the numerals representing n. What can we say to (7.24)? If it is true, then it is disprovable. But since the disproof procedure in question clearly produces only false sentences in the models of T, then (7.24) must be false in such models. Hence it cannot be true in them.

But fairly obviously (7.24) is, like the comparable predicate in ordinary first-order languages, expressible in an ordinary first-order form. Hence its strong and contradictory negations coincide. Therefore there exists a sentence,

$$\sim Disp(\mathbf{n}) \tag{7.25}$$

which is true but whose negation is not disprovable. Hence the theory T is consistent but incomplete in the sense that the negation (7.24) of (7.25) cannot be shown by our disproof procedure to be inconsistent with T.

Thus the existence of a complete disproof procedure does not result in paradoxes but in a remarkable positive result. It shows that if a statement (e.g., a finitely axiomatizable theory) T expressible in IF first-order language is consistent (in the weak sense of not being logically false), its consistency in this sense can be proved by establishing the truth of a certain particular sentence in the same language. In this sense, problems of consistency, like problems of truth, can always be formulated in the same IF language as the original theory T. At the same time, we can see that T is inevitably incomplete deductively: there will be sentences that are false (in the domain of natural numbers) but not disprovable on the basis of T.

The behavior of negation in IF first-order languages is also one of the sources of the limitations of IF first-order languages mentioned in Chapter 4 above. One aspect of these limitations confronts us in the theory of definitions. Indeed, IF first-order logic is likely to put the entire role of definitions in logic and mathematics in a new light. The current view of explicit definitions is that they are noncreative. Whatever can be proved (or defined) by using a definition must be provable (or definable) without using it. Neither of these requirements is satisfied in IF first-order logic. In order to see this, consider an irreducibly IF first-order formula $S(x)$ with x as its only free variable. Suppose we want to abbreviate $S[x]$ by introducing a new one-place predicate constant $P(x)$ as a shorthand. There is of course no harm in doing so, as long as we realize that the constant so introduced is not on a par logically with the primitive nonlogical constants of the underlying language. For such an introduced constant the law of excluded middle does not hold, unlike the original nonlogical constants of the language. From that it follows that the explicit definition of $P(x)$ cannot be expressed by a sentence like

$$(\forall x)(P(x) \leftrightarrow S[x]) \qquad (7.26)$$

for (7.26) implies that the law of excluded middle applies to $S[x]$.

Furthermore, we cannot use the *definiens* $P(x)$ in other formulas that are abbreviated definitionally. Such dependent abbreviative definitions can introduce false assumptions into one's argument.

These remarks suggest a way of extending IF first-order logic in the form it has been defined here. This way consists in allowing the nonlogical constants (predicates and functions) to be only partially defined, so that they need not satisfy the law of excluded middle. This would open interesting possibilities of further developing IF first-order logic as has been explained in this book. Exploiting these possibilities would bring the theory of IF first-order logic in cooperation with the theories of partiality and partial logics. (They are surveyed in a masterly fashion by Fenstad, 1996.) In spite of the great potential interest of such developments, I will not indulge in them here. Instead, I will explain a different kind of extension of the original IF first-order logic of the Chapter 3 vintage.

(C) Contradictio ex machina

At this point of my mystery story of the missing negation, I fully expect Lieutenant Colombo to scratch his head and say to me: "Yes, but there is still one question that bothers me. Why cannot we simply introduce a contradictory negation by a fiat? All we need to give a semantical meaning to such a negation, say $\neg S$, is to stipulate that $\neg S$ is true just in case S is not true. What's wrong with such a procedure?".

As always, Colombo is right. There is nothing wrong introducing a contradictory (weak) negation as it were an afterthought. The result will be called an *extended IF first-order language*. Its logical vocabulary is the same as that of our basic IF first-order language, that is, $\sim$, &, $\vee$, $\exists$, $\forall$ and $=$, plus the contradictory negation $\neg$. These languages have several interesting properties, some of which will be discussed later. Their semantics is determined by the earlier rules of semantical games for IF first-order logic plus a semantical rule for contradictory negation. However, there are conditions which such a semantical rule must satisfy.

(a) The only semantical rule which one can give to the contradictory negation is the negative one:

$\neg S$ is true if and only if S is not true, otherwise false. $\qquad$ (R.$\neg$)

In particular, no game rules can be used to define contradictory negation. One way of seeing this is to point out that the contradictory negation should certainly obey De Morgan's laws, the law of double negation and the usual laws for negated quantifiers (i.e., the equivalence of $\sim (\exists x)$ with $(\forall x) \sim$ and of $\sim (\forall x)$ with $(\exists x) \sim$). But it can easily be seen that if you assume these laws as a part of your system of game rules, then you end up with the strong (dual) negation and not the contradictory one. Hence, in view of these observations, if contradictory negation could be defined by semantical game rules, it would have to coincide with our dual (strong) negation, which in fact it does not.

(b) In the sense that can be seen from (a), contradictory negation is in a game-theoretical treatment of the semantics of IF first-order languages inevitably parasitic on the dual negation.

(c) Another consequence of (a) is that *contradictory negation can syntactically speaking occur only in front of an entire sentence* (closed formula). It cannot be prefixed to an open formula, for then you would need a game rule to handle a substitution-instance of that open formula when you reach it in a semantical game.

There are certain exceptions to this restriction, but they do not change the main point. For one thing, $\neg$ can occur in front of an atomic formula for no further rule applications are needed in that case. Furthermore, one can obviously allow contradictory negation to occur within the scope of &, $\vee$ and $\neg$. Even in the quantificational contexts, the difference between $\sim$ and $\neg$ is sometimes immaterial. Consider, for instance, a game of the form $G((\exists x)S[x])$. The first move here is made by the initial verifier. He (she, it) has to choose a member $b \in \text{do}(\mathbf{M})$ (of the domain of the model in question) such that he (she, it) has a winning strategy in $G(S[b])$. But this means that the initial verifier fails to have a winning strategy available in $G((\exists x)S[x])$ if and only if there is none in any game of the form $G(S[d])$ for any $d \in \text{do}(\mathbf{M})$.

But what this means is that

$$\neg (\exists x)S[x] \tag{7.27}$$

is true if and only if

$$(\forall x)\neg S[x] \tag{7.28}$$

is likewise true, with the initial quantifier $(\forall x)$ being governed by its usual game rule. In other words, the symbol combination $(\exists x)\neg$ always makes perfect sense when it occurs sentence-initially, unlike (as you can easily see) the symbol combination $(\forall x)\neg$.

There is one more exception to the ban on contradictory negation inside formulas. When a negation-sign is prefixed to an atomic formula, it clearly does not matter whether it is viewed as expressing a dual or a contradictory negation. This observation does not matter a great deal for the theory of IF formal languages, but it helps us to understand some aspects of negation in natural languages.

The result of adding the contradictory negation $\neg$ to IF first-order logic creates a type of logic which may at first seem somewhat strange, both syntactically and semantically. Syntactically, because the way $\neg$ can occur is restricted in an unusual way. Semantically, because the semantical rules for $\neg$ is of a kind different from the semantical rules for the other logical constants. Methodologically, it is especially interesting to see the interplay between syntactical and semantical rules here. For instance, if we had tried to introduce $\neg$ in such a way that it could occur in front of any formula or subformula also insider a sentence, we could not have given semantical rules for the kind of negation so introduced. One corollary therefore is that we cannot assume that each logical constant, not even each propositional connective, can be introduced by recursive formation rules of a usual sort, in spite of the glib assumption of some philosophers that we must always be able to do so. It could be argued that this syntactical dogma has been instrumental in smuggling the hopeless semantical assumption of compositionality into the philosophers' and linguists' theories.

(d) Strictly speaking, we thus have two different extensions of IF first-order logic. If we merely add all contradictory negations of IF first-order sentences (closed formulas), we obtain what we will call extended IF logic. If we also add truth-functional combinations of (formulas obtainable by means of &, $\vee$ and $\neg$ from) IF first-order formulas, we obtain what will be called truth-functionally (TF) extended IF first-order logic.

(e) By moving to extended IF first-order logic, one of the unsatisfactory features of its unextended variant can be avoided. In unextended IF logic, no deduction theorem holds. Given two sentences S_1 and S_2, there is in general no sentence S_0 which is valid if and only if S_2 is true in every model in which S_1 is true. For instance, $(S_1 \supset S_2)$ is valid if and only if S_2 is true in each model in which S_1 is not false, which is in general a different matter altogether.

In contrast, in extended IF first-order logic, $\neg S_1 \vee S_2$ fills the bill, being true if and only if S_1 is not true or S_2 is true.

This shows the rationale of the extended IF first-order logic as compared to its unextended version. Looking back at the two main functions of logic that were distinguished from each other in Chapter 1, we can see that the notion of contradictory negation and therefore the extension of IF first-order logic is vital for the deductive function of logic. Without the extension, the deduction theorem does not hold. And its not holding means that an inference from the truth of S_1 to the truth of S_2 is not guaranteed by the truth of any formula of the language in question. Hence one of the purposes of the extension is to facilitate the deductive function of logic.

(f) The introduction of weak (contradictory) negation nevertheless necessitates a review of the basic concepts of IF first-order logic. The reason is that I have so far dealt with two sentences that the *true* in the same models as being logically equivalent, even though they can be *false* in different sets of models. Two sentences that are logically equivalent have equivalent contradictory negations, but not always equivalent strong (dual) negations. Moreover, second-order translations of first-order sentences preserve only truth, contradictory negation, and logical equivalence as it has so far been defined, but not necessarily falsity. Hence two first-order sentences may have the same second-order translations, but different strong negations. Accordingly, two first-order sentences may be logically equivalent but one of them can be impossible to represent as an ordinary first-order sentence in the sense that its strong negation differs from contradictory negation.

All this may at first seem confusing, but there is nothing paradoxical about it. What follows is that we must be careful when using the notion of equivalence as applied in IF first-order logic. It might in fact be appropriate to distinguish the *strong equivalence* of S_1 and S_2, which preserves truth and falsity, from their *weak equivalence*, which is the sense of logical equivalence so far employed and which means only that S_1 and S_2 are *true* in the same models.

One consequence that follows from here is that the logical equivalence of the second-order translations of S_1 and S_2 guarantees only their weak equivalence, but not their strong equivalence. Thus in a sense IF first-order logic seems richer than second-order logic in that it facilitates finer distinctions than the latter. This is only apparent richness, however, for the same distinctions could be made on the second-order level, albeit by using nonstandard notions of negation.

After having explained the idea of extended IF first-order logic, it is time for me to return to the crucial question of truth definitions. We have to ask: Why doesn't the contradictory negation recreate the liar paradox? Indeed, some of the versions of the paradox are formulated precisely in terms of Gödel-type sentences which say, so to speak, "I am not true" rather than "I am false".

But how could we construct such a sentence? Presumably by means of the diagonal lemma. But in order to use the diagonal lemma here, we would have to apply it to an open formula. $\neg T[x]$, where $T[x]$ is the truth predicate. But such a formula is ruled out by the restrictions as to where $\neg$ can occur in well-formed formulas. It cannot occur prefixed to open formulas.

(g) What this means is that it makes sense to apply the notions of truth and falsity also to an extended IF first-order language. It must be noted, however, that extended IF first-order languages do not allow for a truth definition that can be formulated in the same language. For, if we were to do so, we would have to say in that language that, for the Gödel number $g(\neg S)$ the contradictory negation $\neg S$ of any formula S,

$$T(g(\neg S)) \text{ if and only if not } T(g(S)) \tag{7.29}$$

where $g(S)$ is the Gödel number of S. But we cannot replace "not" in (7.29) by $\neg$, because it would then be prefixed to an open formula, and in particular to an open formula in the scope of a sentence-initial universal quantifier.

I do not see that this constitutes an objection to extended IF languages. Just because the notion of contradictory negation is so natural, I do not see that it has to be captured by a truth definition. For another thing, any finite number of sentences of the form

$$T(g(\neg S)) \leftrightarrow \neg T(g(S)) \tag{7.30}$$

can be added to an extended IF first-order language without any problems. Furthermore, in extended IF first-order languages we can have a partial truth definition which applies to all sentences (closed formulas) S that do not contain contradictory negation $\neg$. Since each $\neg$ occurs prefixed to such an S, this truth definition indirectly determines the truth and falsity of all sentences of extended IF first-order languages, and hence does the job of a truth definition as fully as can be expected. In this qualified sense, truth definition is possible also in extended IF first-order languages.

In the sense that appears from these remarks, a truth definition for an unextended IF first-order language indirectly defines the notion of truth also for the corresponding extended IF first-order language.

This is in fact the best one can possibly do here, in the sense that the limitations that rule out a translation of a universal generaliz-ation of (7.29) in one's object language are unavoidable. For if these restrictions were removed, we could formulate a truth predicate for an extended IF first-order language. All we have to do is to add to the conjunction of clauses in the predicate something like the following:

$$(\forall x)(\forall y)((\text{Neg}(x, y) \supset (\neg X(x) \leftrightarrow X(y)))) \tag{7.31}$$

where $\text{Neg}(x, y)$ is the relation that holds between x and y if and only if y is the Gödel number of the contradictory negation of the formula with the Gödel number x. This relation is obviously representable as soon as our language contains a modicum of elementary number theory.

But the extended truth predicate allows for the liar paradox. The paradox is obtained by applying the diagonal lemma to the formula $\neg\, \text{Tr*}(x)$, where $\text{Tr*}(x)$ is the extended truth predicate. The diagonal lemma yields a number n such that

$$g(\neg\, \text{Tr*}(\mathbf{n})) = n \qquad\qquad (7.32)$$

But (7.32) is paradoxical. Hence the restrictions mentioned are unavoidable.

(h) Consider now the fragment of extended IF first-order logic which contains only sentences of the form $\neg\, S$, where S is a sentence of the unextended IF first-order logic. This fragment has several remarkable properties. For instance, we can test whether $\neg\, S$ is logically true by trying to construct a model for S. As was pointed out in Chapter 1, there is a complete set of rules for doing so. Hence, $\neg S$ is logically true (i.e., true in every model of the underlying language) if and only if this construction process comes to a dead end. But this means that the logic of sentences of the form $\neg\, S$ (where S is an IF first-order sentence) is semantically complete.

(i) The results of the last two sections will be put in an interesting perspective by further observations. It was said earlier that IF first-order logic is much stronger than ordinary first-order logic. Extended IF first-order logic is even stronger. How strong? That depends on how you measure strength. In Chapter 9 it will be pointed out that there is a sense in which one can do all of ordinary mathematics by means of the extended IF first-order logic, not just notationally but model-theoretically. Among other things, for each normal mathematical theory there is one in an extended IF first-order language having *mutatis mutandis* the same models.

What such results imply is that extended IF first-order logic is an infinitely better candidate for the role of a mathematical *mathesis universalis* than, for example, set-theoretical languages. The reason is that by giving a characterization of truth for extended IF first-order languages (albeit only via a truth definition for the corresponding unextended IF first-order language) we can in effect fulfill Carnap's dream and give a characterization of mathematical truth (mathematical validity, if you prefer), and to give it in the same language by means of which you can in principle do your mathematics. For any sufficiently rich mathematical theory one can then characterize the concept of truth for it in that very same language. This is something that you just cannot do fully in set theory, as will be shown in Chapter 8.

How important this fact is can be seen by recalling the deductive incompleteness of most mathematical theories. What this means is that an important part of a mathematician's work is the search for new deductive principles suitable for the theory that he or she is pursuing. My results show that if that theory is formulated in an extended IF language, that search can be conducted by means of the very same language. A language like that of set theory which does not allow the formulation of descriptively complete theories does not enjoy a comparable privilege. Thus my results have important consequences for the formulations of mathematics in general.

As an example of these consequences, it may be pointed out that in extended IF logic one can formulate a descriptively complete axiom system for elementary number theory. In order to do so, one basically needs, over and above the usual first-order Peano axioms, an axiom of induction as strong as its second-order version

$$(\forall X)((X(0) \mathbin{\&} (\forall y)(X(y) \supset X(y+1)) \supset (\forall y)X(y) \tag{7.33}$$

But (7.33) is a Π_1^1 sentence, hence the contradictory negation of a Σ_1^1 sentence. Since this Σ_1^1 sentence has an equivalent IF first-order translation, (7.33) has an equivalent in the corresponding extended IF first-order language. And this implies that elementary number theory admits of a descriptively complete axiomatization by means of extended IF first-order logic.

(D) Negation is a Siamese Twin concept

Our results put the entire concept of negation within a surprisingly new light. Logicians and philosophers sometimes think of negation as a simple and unproblematic idea. All that is involved, so it might seem, is an inversion of truth-values. For instance, in the *Tractatus* Wittgenstein held that the negation of a pictorially interpreted sentence is not only also a picture, but the same picture, only with its polarity reversed. Perhaps logicians and philosophers should have taken a cue from linguists, who have found ordinary-language negation a complex and puzzling phenomenon.

Our results lead to a striking conclusion. What they show is that *in any sufficiently rich language, there will be two different notions of negation present.* Or if you prefer a different formulation, our ordinary concept of negation is intrinsically ambiguous. The reason is that one of the central things we certainly want to express in our language is the contradictory negation. But it was found earlier in this chapter (cf. (a)–(b) above) that a contradictory negation is not self-sufficient. In order to have actual rules for dealing with negation, one must also have the dual negation present, however implicitly.

What is truly remarkable here is not just the possibility of having two different notions of negation present in one's language, but the virtual necessity of doing so in sufficiently strong languages. The crucial fact here is not that we can introduce a contradictory negation in the way we have done. Rather, it is the fact that it *must* be so introduced. The contradictory negation cannot be introduced by means of any game rules. Hence it has to be different from the logically primary (dual) negation that can be defined by the game rules.

Another important fact is that the strong (dual) negation can be introduced in only one way. Its properties are completely fixed in an IF first-order language.

What this implies for the semantics of natural languages is clear. We might as a thought-experiment think of the language community as if it were facing the task of creating a sufficiently strong artificial language for roughly the same purpose as a natural language. Such an imaginary legislative linguist faces an interesting problem. On the one hand, the most important thing that has to be expressed in such a language is clearly the contradictory negation. But it cannot be the only notion of negation present. Something else has to be involved in the language, however tacitly. For we just cannot formulate adequate semantical or syntactical rules for the language in question in terms of the contradictory negation alone. Without the dual negation, one's language is just not adequate for the purposes of logical and semantical processing. (For instance, the rules of semantical games have to be formulated in terms of the dual negation.) If our imaginary linguistic inventor uses only one symbol for negation, then he or she has to put that one symbol to two essentially different uses. This inevitably leads to complications of some sort or the other.

Now the language community is in an essentially similar position as the imaginary language designer, for in natural languages we seem to be dealing with one negation only, namely, the contradictory one. Accordingly, the behavior of negation in natural languages is likely to exhibit curiosities caused by the need of also using, however tacitly, the other (dual) negation. This suggests an interesting perspective from which to view the phenomenon of negation in natural languages.

I cannot attempt here a full-scale discussion of negation in natural languages. I am in any case fully convinced that the results reached in this chapter will also be an essential part of any half-way satisfactory theory of negation when it comes to natural languages. For one thing, in spite of the confusing complexities of natural languages, several features of the behavior of negation in English become more understandable when they are compared with what we find in extended IF languages.

Among other things, we obtain an interesting perspective on the bifurcation of natural-language negation into verbal and sentential negation. It would be a serious oversimplification simply to assimilate this dichotomy to the distinction between strong and weak negation in extended IF languages; but the two contrasts are not unrelated, either. When a negation-sign occurs in front of an atomic formula in a formal language, the difference between strong (falsifying) negation and weak (contradictory) negation disappears. Hence verbal negation, which very roughly speaking negates the unquantified part of a sentence, can be construed as the contradictory negation which the natural language negative ingredient obviously expresses in the first place.

At the same time, the vagaries of contradictory negation in natural languages like English become understandable. The basic fact is that explicit ordinary-language negation is the contradictory one. Complications arise partly because such a negation is not self-sufficient, and partly because it cannot always occur meaningfully in all contexts in which it should, purely syntactically speaking, be admissible. Hence it may be expected that there are semantical regularities which semantically speaking move either to a sentence-initial position or to a position which corresponds to fronting an atomic formula. For instance, the fact that we can easily form the contradictory of an English sentence by prefixing it by "It is not the case that" is like introducing a sentence-initial contradictory negation into an IF first-order logic without being able to introduce it into any position except the sentence-initial one. The difficulty or perhaps even impossibility of giving effectively applicable rules for forming contradictory negations in English is in keeping with the inevitable absence of contradictory negation from an entire IF first-order language, without extending it by a new type of formation rules and by a new type of truth-condition.

There even seems to be linguistic regularities in English which are closely related to the awkwardness of locating negation between a sentence-initial position and a minimal-scope position in an English sentence, apart from truth-functional compounding. One regularity is seen by raising the following question: When can *not* be prefixed to a quantifier phrase in English? The following examples illustrate some of the acceptable constructions:

> Not every Scotsman is stingy.   (7.34)
> Not a single student failed.
> Not many runners managed to finish.

In contrast, the following constructions are unacceptable:

> not any   (7.35)
> not each
> not some
> not few
> not several

What makes the difference? It turns out that the quantifiers involved in (7.35) all have a wider scope than negation while those in (7.34) do not. Hence the constructions (7.35), if they were admissible, could push negation into the sentence-internal no-man's-land. In contrast, the quantifiers occurring in (7.34) do not have the right of way in relation to negation. Therefore they let negation stay sentence-initial, and hence presumably a contradictory negation.

This explanation is reinforced by observing that the constructions instantiated in (7.34) are *ceteris paribus* admissible only in the subject position, but not in the object position. For instance, one can say

> Not every hunter shot a lion.   (7.36)

but not

> A hunter shot not every lion.   (7.37)

The reason is that if *not* occurs in a phrase which occupies an object position, it can be within the scope of other quantifiers, typically quantifiers in the subject-position phrase. This would once again land negation in a position where contradictory negation cannot occur, if we assume that the English is like extended IF language.

This regularity has nothing to do with the "intended" meaning of expressions like (7.37), for it is acceptable to say

Not every lion was shot by a hunter. (7.38)

There are other linguistic regularities which become understandable in the light of the behavior of negation in IF languages. One of them is the fact that overt negation can be a barrier to anaphora. This fact has been noted in the literature, and Irene Heim (1982, pp. 115–117) has even used its alleged inexplicability in game-theoretical semantics as a basis for an attempted criticism of GTS. Now it can be seen that such attempted objections can be turned into evidence for game-theoretical semantics, and not against it.

This phenomenon is exemplified by the fact that anaphora is possible in the former of the following two sentences, but not in the latter:

Some student passed the examination. She must have studied
very hard. (7.39)

Not every student failed to pass the examination. She must
have studied very hard. (7.40)

Since (7.39) and (7.40) are logically equivalent, it might at first seem puzzling that anaphora is acceptable in one of them but not in the other. This fact is nevertheless easily accounted for. In Hintikka and Kulas (1985), a theory of anaphora is developed within GTS. Apart from the details the basic idea is clear, and it is the only thing I need to evoke here. The idea is that, in any play of a semantical game, anaphoric pronouns refer to certain individuals introduced earlier in the course of the same play. For instance in (7.39), the first game rule to be applied is the rule for *some*. It involves a choice of an individual, say Susan, from the relevant domain, whereupon the game is continued with respect to

Susan passed the examination. Susan is a student. She must
have studied very hard. (7.41)

When some time later in the game the pronoun *She* is dealt with, there will be an eligible value for it available, namely, Susan. Hence anaphora is possible in (7.39).

In contrast, (7.40) is not interpreted directly through any game played with it. Since its first sentence obviously involves a (sentence-initial) contradictory negation, it is interpreted only indirectly as the contradictory negation of

> Every student failed to pass the examination. (7.42)

Even if a game with (7.42) is involved in the interpretation of (7.40), it does not automatically restore the possibility of anaphora in (7.40). There are two concurrent reasons for this. On the one hand, what the first sentence of (7.40) says is that there is no winning strategy for a game with (7.42). Hence it does not give any clues as to what kind of play with (7.42) might be relevant here, and consequently what individuals might be chosen by the players in such a game. Therefore there are no well-defined individuals in the offing to serve as the value of *every* in (7.40) and hence as a value of the anaphoric pronoun *She*.

On the other hand, individuals introduced by applications of rules for universal quantifiers (like the *every* in (7.42)) are not automatically available as values of anaphoric pronouns. (I will not discuss here the reasons and limits for this regularity.)

In any case, this regularity has nothing to do with the quantifier *every* as such. For we have a similar situation with pairs of sentences like the following:

> Few students passed the examination. They must have
> studied very hard to be able to do so. (7.43)

> Not many students passed the examination. They must
> have studied very hard to be able to do so. (7.44)

There are prima facie exceptions to the rule that negation is a barrier to anaphora. They are nevertheless precisely the kinds of exceptions that literally prove the rule. In fact, they are examples in which the semantical game with the unnegated sentence is brought in and considered explicitly or implicitly, usually to provide a reason why the unnegated sentence is false. The following is a case in point:

> Nobody stole your diamonds, for he would have had
> to scale a ten-foot wall to do it. (7.45)

An even simpler regularity which becomes understandable from our vantage point is the semantical ordering principle that gives negation priority over *every* also when the negation ingredient is within the syntactical scope of *every*. For instance,

Everything that glitters is not gold. (7.46)

does not have the logical form

$(\forall x)(x \text{ glitters} \supset \neg(x \text{ is gold}))$ (7.47)

but rather

$\neg(\forall x)(x \text{ glitters} \supset x \text{ is gold})$ (7.48)

Why? Because the former reading would place a contradictory negation in the scope of a universal quantifier where it is not interpretable in all similar sentences. What is interesting about this phenomenon is that it occurs in a wide variety of languages, some of them completely unrelated to each other. It therefore cries out for a theoretical explanation. The results reached in this chapter provide such an explanation.

This list of explanations of natural language phenomena could be continued. Similar problems have in fact been noted by linguists, for instance in connection with the phenomena of neg-raising (cf. e.g., Horn 1989, pp. 308–330). Many of them admit of explanations along the same lines as the phenomena noted here. What has been said suffices to show that IF first-order logic will play a crucial role in such a study.

It is to be noted also that all we need to solve the informal liar-type paradoxes is to acknowledge that negation behaves in natural languages in essentially the same way as in extended IF first-order languages. An analysis of the self-referential mechanism which leads to such prima facie paradoxes is needed only for the purpose of arguing for the naturalness of this solution, and not for the solution itself. For instance, consider the archetypal liar sentence

Sentence (7.49) is false. (7.49)

What kind of procedure is involved in trying to verify (7.49)? Such a language game would have to involve some rule like the following:

When the game has reached a sentence of the form, "Sentence **(n)** is false", the verifier has to look up sentence #n. The game is then continued with respect to its negation. (7.50)

A mirror-image rule would be

> When the game has reached a sentence of the form, "Sentence
> (n) is true", the falsifier has to look up sentence #n. The
> game is then continued with respect to it.   (7.51)

Applied to (7.49), a game with the rule (7.50) leads to a loop, and hence to an infinite play of the game. This does not by itself prevent us from speaking of winning and losing here. We know from game theory that there can be perfectly reasonable ways of defining who wins and who loses in infinite and not only finite games. However, in this particular case, the infinite play is symmetrical with respect to the two players. Hence there is no reasonable rule for winning or losing that would declare either player the winner. Hence (7.49) cannot be verified, and by symmetry it cannot be falsified, either. Therefore (7.49) cannot naturally be considered either true or false.

A similar line of thought can be applied to

> The sentence (7.52) is true.   (7.52)

In this case, however, the infinite play of the game is not symmetrical. One of the players, namely, nature (the initial falsifier) can be "blamed" for the play's infinitude. This can naturally be used as a basis for stipulating a sentence like (7.52) to be true. The idea is that if one player bears the whole responsibility for the play going on to infinity, then the blameless one wins (cf. here Hintikka and Rantala, 1976).

Hence an attempt to vindicate the strengthened liar paradox by considering a sentence like

> The sentence (7.53) is not true.   (7.53)

where "not true" expresses contradictory negation also fails, for (7.53) is simply false.

You do not have to like the details of this treatment of informal liar-type paradoxes. For even though this solution might very well look like an ad hoc procedure, it in any case serves to illustrate the ease with which liar-type paradoxes can be dealt with in natural languages as soon as it is admitted that negation behaves in natural languages basically in the same way as in extended IF first-order languages.

The kinds of definitions of winning and losing and a fortiori of truth and falsity are interesting in principle (among other things) because they show that there are reasonable ways of defining truth and falsity also for what Kripke calls ungrounded sentences containing the truth predicate (see Kripke in Martin 1984, p. 57).

(j) Among the other features of the concept of negation that have to be deconstructed is the so-called law of excluded middle. Since there must be two different concepts of negation in a sufficiently strong language, the question of the validity of the law of excluded middle is intrinsically and inevitably ambiguous between the two negations.

Moreover, for each of the two negations, the corresponding question concerning the *tertium non datur* admits of a simple answer. For the contradictory negation, the law of excluded middle holds virtually *per definitionem*. For the dual (strong) negation, it fails almost equally trivially. What is particularly remarkable is that both answers represent straightforward logico-combinatorial facts which are not tied to any particular epistemological or other philosophical assumptions.

One remarkable thing about the failure of *tertium non datur* in IF first-order logic is that this logic is in some sense perfectly classical. The failure is a consequence of changes in the ordinary first-order logic which were motivated completely independently of all "nonclassical" ideas, such as constructivism or intuitionism. Hence my results put the entire contrast between classical and nonclassical logic in a new light. Or, rather, they confound it totally by showing that the received logic (received ultimately from Frege and Russell) does not deserve the epithet "classical" except in the sense of being the logic that is taught in classrooms these days.

This point is best explored in connection with the reconstructions of constructivistic ideas discussed in Chapters 10 and 11.

(k) In a truth-functionally extended IF first-order language, we can finally express normal truth-functional conditionals, for example,

$$(\neg S_1 \vee S_2) \tag{7.54}$$

which can be abbreviated

$$(S_1 \supset_T S_2) \tag{7.55}$$

Hence a truth-functionally extended IF first-order language is a happier medium of actually proving propositions logically than the unextended IF first-order logic. This gain in syntactical manipulability is nevertheless inversely related to the perspicuity of the semantical situation in the two types of languages. In the original model-T variant of IF first-order languages we did not have truth-functional conditionals, which makes a proof theorist's life difficult, but we have all sorts of nice model-theoretical results. In contrast, in truth-functionally extended IF first-order languages we do have truth-functional connectives, ordinary *modus ponens* and so forth, making chains of actual logical inferences much easier. But that means being banished from the model-theoretical Eden of the original plain IF first-order logic. There is a price to be paid here, no matter which way we go. A main restriction that must be observed here is that the truth-functional conditional $\supset_T$ can only be used outside quantifiers, like the truth-functional negation. Fortunately, this presents no new obstacles to the use of conditionals in actual logical inferences.

(1) My results admit of a striking formulation in Wittgensteinian terms. Wittgenstein spoke of the limits of language. Usually, the limits of what a formal language can express are thought of in terms of limitations to the expressive power of the language or perhaps in terms of limitations of the totality of entities it can deal with. What we have found shows that one of the true limits of language lies in the inexpressibility of contradictory negation or, if such a negation is introduced by a *fiat*, in the consequences of that forced expressibility. For instance, the crucial but inevitable restriction on IF first-order logic is the inexpressibility of contradictory negation, outside the ordinary first-order fragment. And the introduction of contradictory negation destroys most of the nice metalogical properties of IF first-order logic, including the definability of truth in the language itself.

8

Axiomatic Set Theory:
Fraenkelstein's Monster?

The backbone of the approach to the foundations of mathematics developed in this book is game-theoretical semantics and the new logic it has inspired, the independence-friendly logic. In the preceding chapters, GTS has opened new perspectives in several directions. IF logic was seen to put the whole array of questions concerning completeness and incompleteness in logic as well as in mathematics in a new light. It frees the model theory of the fetters of set theory by facilitating the formulation of a truth predicate for a suitable IF first-order language in the language itself. Furthermore, in IF first-order languages the important idea of negation turns out to behave in an unexpected way, thereby suggesting that the same might be true of natural languages. But even after all these results many important questions still remain unanswered, including comparative ones. IF logic may have some advantages as a framework of mathematical theorizing, but what has been seen scarcely seems to affect the virtual monopoly of set theory in that department. Admittedly, there is no longer any need to resort to set theory or even to higher-order logic for the purpose of doing model theory of first-order logic. But this relative self-sufficiency of IF first-order languages does not prove their superiority over set-theoretical languages. On the contrary, set theory is still the medium of choice for many mathematicians, philosophers and logicians for model-theoretical theorizing. In this chapter, it will be argued that this privileged role of axiomatic first-order set theory becomes extremely dubious in the light of the insights that the game-theoretical approach has yielded, and will yield.

It seems to me that history is very much on my side. This role of set theory as universal mathematics goes back to the founding fathers of set theory. Cantor thought that any structure could in principle be dealt with in set theory (see Dauben 1979, pp. 229–230). This was not his private theory, either. Set theory was thought of as providing models for all possible mathematical theories. One context in which there is a need for different kinds of models is created by various consistency proofs. Beltrami and Felix Klein had impressed the mathematical world by presenting a consistency proof for an axiomatic non-Euclidean plane geometry relative to the Euclidean one. They had constructed a model for a nonEuclidean geometry within a model of the Euclidean one (cf. Torretti 1978, pp. 115–142). But other varieties of non-Euclidean geometry did not admit such models. Where, then, can we find the models that serve to show the consistency of such geometry? In set theory, it was thought. This is the reason why Cantorian set theory seemed a veritable paradise to David Hilbert (1925, p. 170) who was looking for consistency proofs for all and sundry mathematical theories. For Aristotle, the hand was the tool of all tools and the soul the form of all forms. In a similar manner, for the creators of set theory this theory was calculated to give us the model of all models. Indeed, in Hilbert's thinking, besides the universe of discourse of some one mathematical theory which we might be considering, we also have to keep an eye on a more extensive domain, namely, the universe of all possible structures. When Hilbert sometimes seemed to infer existence from consistency, incidentally upsetting poor Frege no end, he meant existence in the super-universe of all structures.[1]

In the course of the subsequent development of axiomatic set theory, this idea of the universality of the universe of set theory was nevertheless lost. The genesis of axiomatic set theory was in effect an attempt to make sure, not that the universe of set theory contains all possible structures but only that it contains all structures representable in the language of set theory. In the simplest case, this means trying to capture in the universe of one's set theory all sets (that is, all subsets of the domain of individuals) definable in the set theory in question rather than all sets, period. This is illustrated by the role of the comprehension axiom or comprehension schema in the axiomatic version of set theory. This axiom schema can be expressed in its unrestricted form as

$$(\exists y)(\forall x)(x \in y \leftrightarrow S[x]) \tag{8.1}$$

where $S[x]$ is any well-formed formula in the language of set theory with x as its only free variable. The paradoxes forced set theorists to give up the unrestricted form of (8.1). It can nevertheless be said that set theorists have tried to capture as many sets as possible by means of (suitably restricted forms of) the comprehension schema (8.1). They have tried to impose as few restrictions on it as is deemed safe. The scheme (8.1) has thus been their main tool in investigating the difficult problems of set existence. In typical axiom systems of set theory, all axioms of set existence are special cases of (8.1), with one exception. This exception is the axiom of choice, which has been argued (and will be argued) in this book to be unproblematic. Yet, even if we could look away from the restrictions that are needed in (8.1), we cannot capture by its means all the sets we are considering – that is, all subsets of the domain of individuals of the intended model of axiomatic set theory. Thus the paradoxes of set theory which have necessitated restrictions in the comprehension axiom are only a part – and the easier part – of the problem of constructing an adequate set theory. The more difficult part is to capture the sets not representable by explicit formulas of one's set-theoretical language.

There is a sense in which the use of the comprehension axiom (8.1) is based on a fallacy, or at least on dated assumptions. The richness that the universe of set theory was originally thought of as having was model-theoretical in nature, as was seen earlier. In contrast, the comprehension axiom (8.1) is an axiomatic and syntactic device to enforce as much richness in the universe of one's set theory as possible. It is perhaps understandable that Zermelo should have thought in 1908 that he could enforce model-theoretical plenitude by axiomatic means. But no one who has internalized Gödel's incompleteness results can reasonably hope to do so any longer.

This matter can be looked at from another point of view. The contrast we have found between the original Cantorian "naive" set theory and axiomatic set theory is clearly related to the distinction between the standard and nonstandard interpretation of higher-order logic. This distinction was first given these labels by Henkin (1950) even though some earlier thinkers, especially Frank Ramsey (1925), had been aware of it and though it had played an important role in earlier foundational discussions (see Hintikka 1995a). It affects all higher-order quantifiers, but it can be explained by reference to a single second-order one-place predicate variable (set variable), say X. What are the (extensions of the) values of such a variable? The standard interpretation says that they are all the subsets of the domain of individuals do(M) of the model **M** in question – that is, that the range of X is the power set of do(**M**). If, instead of this interpretation, we restrict the values of X to some designated subset of the power set of do(**M**), we obtain a nonstandard interpretation of the variable X. The most common restriction is to the sets definable to the theory in question.

A similar distinction can be applied to any higher-order variable. It is usually required that the set of designated sets (of any logical type) is closed with respect to Boolean operations and projective (quantificational) ones. When I speak of the descriptive completeness of axiomatic set theory, it is of course by reference to some kind of standard interpretation. However, the analogy between the different interpretations of set theory and the different interpretations (i.e., the standard vs. nonstandard contrast) of higher-order logic is not perfect. The standard versus nonstandard distinction can be only locally applied to models of axiomatic set theory. We can speak of standardness locally, for instance, in connection with a given set, by reference to the question as to whether all its (extensionally possible) subsets are represented in some given model of one's axiom system for set theory, or to the question whether the set of natural numbers as reconstructed in axiomatic set theory contains only "standard" natural numbers. In contrast, the entire set-theoretical universe is inevitably nonstandard. Either the set of all sets or its power set fails to exist, though both are extensionally possible and hence ought to exist in a standard model.

What was pointed out is that an ordinary first-order axiomatic set theory cannot guarantee full standardness. In this sense, axiomatic set theory fails in its attempt to recover the Cantorian paradise. Different versions of the axiom of comprehension inevitably fail to do their intended job of capturing all the sets that we would like to include in the universe of set theory, not because they are not the right formulations of the axiom, but because they are first-order axioms. As a consequence, axiomatic set theory is inevitably incomplete descriptively. And since it uses first-order logic, which is semantically complete, set theory is also deductively incomplete. (If you have not thought of this incompleteness before, recall that you can do elementary number theory by means of axiomatic set theory, and then apply Gödel's incompleteness theorem.) The crucial question is: How serious is this failure? A related question that arises here is: Why bother with an unavoidably incomplete theory? The most respectable prima facie reason that I can see to cultivate incomplete axiomatic set theory is to think of it as an approximation to the real thing, an approximation which can be made closer and closer by the addition of new axioms. This way of thinking nevertheless involves a serious mistake. To speak of approximations presupposes that there is something to approximate. And the lack of a standard model of axiomatic set theory means that there is apparently nothing to approximate. We have no unproblematically clear idea of what the intended models of set theory are. Hence it makes no sense to speak of additional axioms that capture the structure of the intended model of set theory more and more closely.

This point is worth spelling out more fully. A comparison between set theory and normal mathematical theories may be instructive. When it comes to many other theories we often have a good idea of what the relevant models are – for instance, what the structures are that are exhibited by natural numbers, by real numbers, by the second number class, and so forth. We know what the different structures are that we want to study in group theory, lattice theory or topology. In contrast, we do not have a sharp idea of what the model of set theory intended by Cantor and his contemporaries is like. This uncertainty is partly a consequence of the paradoxes that arose in early set theories. It is usually formulated by speaking of the difficulty of deciding which axioms we should adopt in an axiomatic set theory. This diagnosis nevertheless deals with the symptoms of the problem rather than with their cause. There cannot exist a complete first-order axiom system for set theory, anyway. Hence the focus of attention should be on the principles of looking for new axioms rather than the justification of any particular candidates. For instance, the idea that set-theoretical axioms can be compared with each other in terms of the "naturalness" of their consequences is theoretically very shallow. How do the people who advocate this idea know that naturalness is not in the eye of the set theorist, and insensitive to the model-theoretical realities of the situation? Instead of relying on such intuitions, we must realize that to have some idea of what axioms are supposed to hold in set theory is in effect to know something about what the intended models of set theory are like. And such knowledge is precisely what philosophers do not have, at least not with sufficient clarity.

One reply that I will inevitably receive is that the cumulative hierarchy of sets provides the intended model I claim to be missing from the foundations of set theory (for able expositions of this conception of set, see Boolos 1971 and Parsons 1977). According to this idea, the universe of set theory is built up by a step-by-step (but infinite) process. At each stage, the new sets that are introduced are all possible collections of sets already reached.

The rejoinder to this objection is that the cumulative hierarchy is not by any stretch of the imagination what the architects of set theory had in mind. Their idea of the set-theoretical universe is that it is the richest possible one. Whatever can exist does exist in the rich and plentiful world of set theory. Like the Lovejoyful god who is so generous that he grants the gift of existence to everything that can possibly exist, a Cantorian set theorist is expected to assume that all possible sets are present in his universe of all sets (cf. Dauben 1979, pp. 225–236; Lovejoy 1936).

Yet the cumulative hierarchy is ontologically stingy. It may be compared to Gödel's constructive model of set theory (Gödel 1940) even though it is not identical with Gödel's. And the main virtue of Gödel's model is ontological economy. All sets existing in that model are obtained by a narrowly defined construction process, which can be described in set theory itself. The famous axiom of constructibility says that this process exhausts the entire universe of sets. The procedure for building up the constructive hierarchy is a more generous one, but it, too, rules out a tremendous variety of potential sets (and their structures). As Hallett (1984, Ch. 6, esp. p. 223) points out very clearly, even though the defenders of the cumulative hierarchy use expressions like "all possible sets", their force is not the same as in ontologically more generous and more realistic approaches to set existence.

Boolos uses the same term "all possible collections" [as Wang and Shoenfield] in his account of the intuitive hierarchy. If we are really *forming* sets in stages, then it may be that at stage *s* we do form all *possible* sets of objects formed at previous stages. But here "all possible" will not be very many, certainly not many compared with the "all possible" of the later "formal" (impredicative) stage of the theory (Hallett 1984).

In general, it is hard to reconcile an emphasis on predicativity or even on any stage-by-stage generation of sets with the Cantorian idea of set theory as the study of all possible structures. Among those possible structures there are very large sets. The iterative (cumulative) conception of sets does not capture them automatically. They can only be captured by special axioms of large cardinals (cf. Kanamori and Magidor 1978). As Gödel emphasizes (1947, reprint p. 476), beyond a certain point the further iteration of the "set of" operation depends on assumptions of large cardinals. The need of such axioms shows that the iterative conception of set is not self-explanatory and certainly does not alone capture the notion of set that Cantor was trying to implement. Even if you do not actually need impredicative sets for certain specific purposes in set theory or elsewhere in mathematics, it is still hard to deny that there are possible structures that can only be characterized impredicatively.

Even without much detailed argument, it is thus clear that the iterative model is not what the founding fathers of set theory had in mind or that such a model of set theory is especially well-suited for the purpose of serving as a framework of mathematics, including the search for new axioms. Such a search is best given a firm direction by some sufficiently explicit characterization of the intended model, which is just what has not been provided by set theorists.

Other candidates for the role of the intended model of set theory will be examined later and found wanting.

In discussing the foundations of set theory, philosophers and mathematicians often talk about the search for new axioms. But such talk is misguided, for nobody knows what one is supposed to look for. In order to illustrate the axiomatic and hypothetical approach to mathematical theorizing, Bertrand Russell once called a mathematician a chap who never knows what he is talking about. In the same spirit of friendly teasing, a set theorist would have to be called a guy (or a gal) who never knows what he or she is looking for.

In brief, axiomatic set theory does not have a model theory which would enable us to characterize its intended models. The deeper reason for this need of a solid model-theoretical foundation of whatever logic or theory that we are envisaging as the ultimate framework of mathematical theories, is seen from the conclusions reached in Chapter 5. There it was seen that the first and foremost goal of mathematical theorizing is a descriptively complete theory, and not a deductively complete one, as many philosophers would like to have it. Now, the notion of descriptive completeness is a model-theoretical one, presupposing both a truth definition and a notion of intended model.

But the difficulties with axiomatic set theory cut deeper than our ignorance (or indecision) of what its intended models are like. We know that whatever the details of the standard models of set theory are or may be, these models cannot be fully captured by an axiomatic set theory formulated on the first-order level. First-order axiomatic set theories are not only deductively incomplete as they stand; they are inevitably incomplete descriptively, no matter what new axioms are added to them. One way of seeing this is to go back to the basic question: Why should we cultivate set theory? The reason why we were first led to consider set theory in the line of thought presented in this book was that set theory (or its equivalent) was apparently needed for the model theory of our basic logic, that is, ordinary first-order logic. This motivation for taking up set theory has been widespread among philosophically oriented logicians, and probably still is. Often it takes a more general form. One of the main jobs that set theory is drafted to do is to enable us to deal with model-theoretical questions in different parts of logic. And this conception of the job that set theory can do is not restricted to logic. The so-called structuralist philosophy of science is little more than an application of this idea (or presumption) to the epistemology of scientific knowledge.

This alleged rationale of set theory is not a genuine one. There is in the literature a widespread tendency to give set theory the credit for theories which have been developed by using set theory as their metatheory. For instance, in Moore (1994, p. 635) we read:

Set theory influenced logic, both through its semantics, by expanding the possible models of various theories and by the formal definition of a model; and through its syntax, by allowing for logical languages in which formulas can be infinite in length or in which the number of symbols is uncountable.

The tacit syllogism on which such claims rest is something like this: logical languages which cannot be dealt with syntactically must be dealt with model-theoretically. Now set theory is the natural medium of model-theoretical conceptualizations. Hence it is by means of set theory that theories like infinitary logics have been developed and are being cultivated. The fallacy here is that nothing is said or done to exclude other possible metalanguages and metatheories. And what has been seen in this book makes this question a most pertinent one. By showing in Chapter 6 how a truth predicate can be defined for a suitable IF first-order language I removed the main obstacle from the path of a model theory of IF first-order logic done in IF first-order logic itself.

The table can in fact be turned on set theory here. Much more can be said than merely to remove one of the many possible reasons for cultivating set theory axiomatically, or in some other rigorous fashion needed in a genuine model theory. First-order logic was originally suspect because on apparently could do its model theory in first-order logic itself. If this was a legitimate prima facie complaint, then it is also fair game to ask whether the model theory of axiomatic set theory can be done in axiomatic set theory itself. This question leads us to consider the possibility of defining a truth predicate for axiomatic set theory in that very same theory. For without a truth predicate, there is little hope of developing anything like a viable model theory. Furthermore, this question is made especially pertinent by the reputed role of set theory as the medium (or at least as a surrogate) of model theory.

It is in fact possible to bring the results reached in Chapter 6 to bear on this task of doing model theory of axiomatic set theory in the same theory. Indeed, axiomatic set theory is an ordinary first-order theory with a single nonlogical predicate containing elementary arithmetic. Hence a truth predicate can be formulated for it in the Σ_1^1 fragment of the corresponding second-order language, that is, in the form

$$(\exists f)\text{Sat}[x, f] \tag{8.2}$$

But what (8.2) expresses in a sense should be expressible also in the first-order set theory itself. Indeed (8.2) is equivalent to a statement of the form

$$(\exists X)S[X] \tag{8.3}$$

where X is a one-place second-order predicate variable. From (8.3) we can obtain a related but not always logically equivalent (first-order) set-theoretical statement by replacing every subformula of (8.3) of the form $X(y)$ (or $X(a)$) by $y \in x$ (or $a \in x$) and by replacing $(\exists X)$ simply by $(\exists x)$. In this sense, we can in fact formulate a truth predicate for first-order axiomatic set theory in the same set theory itself.

But this observation needs to be put in perspective. At first sight, it seems to violate Tarski's impossibility result according to which the concept of truth in a first-order theory cannot be defined in that theory itself.

A beginning of a solution to this puzzle lies in the fact that no truth definition has been given for axiomatic set theory in that theory itself – only a truth predicate. And from the mere formulation of a complex predicate no contradiction can follow.

But in what sense, if any, does Tarski's result then rule out the possibility of self-applicable truth predicates? In order to see what this sense is, the reader is invited to go back to the argument at the beginning of Chapter 7 which served to show that the truth predicate (8.1) considered there leads to a contradiction. We implicitly used there the assumption that the sentence $\sim T[\mathbf{n}]$ constructed by means of the diagonal lemma is logically equivalent to (or at least has the same truth-value in the intended model as) the arithmetical sentence that says that the sentence with its very own Gödel number $g(\sim T[\mathbf{n}]) = n$ is true. But $\sim T[\mathbf{n}]$ says that the same sentence is false; hence a contradiction which can be taken to have the form

$$\sim T[\mathbf{n}] \leftrightarrow T[g(\sim T[\mathbf{n}])] \tag{8.4}$$

where the right side is of course equivalent to $T[\mathbf{n}]$.

The equation of the truth-values of the two sentences $\sim T[\mathbf{n}]$ and $T[g(\sim T[\mathbf{n}])]$ is a consequence of the adequacy conditions on any candidate for the role of a truth predicate. The operative adequacy condition is, to all practical purposes, Tarski's familiar T-schema (see Chapter 6) applied to the game-theoretical truth predicate. Both are incomplete, but neither leads to a contradiction.

This point can be generalized. A truth predicate has consequences only in conjunction with an adequacy condition which is essentially tantamount to Tarski's T-schema. This condition requires that any sentence S is true if and only if the sentence $T(g(S))$ is true. The latter sentence says of course that the truth predicate applies to the Gödel number $g(S)$ of S.

The truth predicate defined in Chapter 6 for suitable IF first-order languages thus accomplished something that Tarski's impossibility result at first sight discourages us from expecting. This something is a truth predicate $T[x]$ which applies to $g(S)$ if and only if S is true according to the game-theoretical conception of truth – which I have argued is our natural conception of truth. The reason why we cannot express this as a formal equivalence

$$S \leftrightarrow T[g(S)] \tag{8.5}$$

is that (8.5) says too much. It asserts also that S and $T[g(S)]$ are false simultaneously. This unwanted implication cannot be avoided in (unextended) IF first-order languages. In such languages the bite of a truth predicate thus lies in its satisfying the adequacy requirement that motivates Tarski's T-schema. However, this requirement cannot be formulated in unextended IF first-order languages as a formal equivalence schema. In this way, new light is thrown both on Tarski's intentions and on their implementation.

What does follow from all this when applied to axiomatic set theory? The first and foremost point here is that, contrary views notwithstanding (cf. McGee 1991, pp. 75–76), there is nothing wrong as such with the truth predicate for axiomatic set theory which was explained earlier in this chapter, and which is formulated in axiomatic set theory itself. In fact, by means of this truth predicate one can do a fair amount of model theory of set theory in set theory itself. We can for instance define what it means for an element of the set-theoretical universe to satisfy a given formula with one free variable. We can then consider the totality of elements satisfying the formula, and so on. This is the objective basis for the surprisingly popular casting of set theory in the role of a do-it-yourself model theory. The implicit idea undoubtedly is in a sense one of economy. Who needs fancy model theory when there already exists a set theory serving as the universal framework of mathematical, and even scientific, theorizing and allowing us to do its model theory in the set theory itself? In fact, much of the actual work in basic model theory of axiomatic set theory can, in this spirit, be thought of as being done by means of set theory itself. Even though such model theory of set theory is ordinarily done informally, it is not hard to believe that much of it can also be done in terms of a suitable axiomatic set theory.

What such a self-applied theory for axiomatic set theory cannot assume are all the instances of Tarski's T-schema, applied to the game-theoretical truth predicate. This follows from Tarski's very own impossibility result. Alternatively, and more directly, we can use the same kind of constructive argument as is usually employed in proving Gödel's first incompleteness theorem by means of the diagonal lemma. What we need to do is to apply this lemma to the negation of the quasi-predicate of truth that I showed above how to formulate for axiomatic set theory in set theory itself. Let this predicate be $T[x]$ and let $g(S)$ be the Gödel number of S. Then by tl diagonal lemma there is a number n such that

$$g(\sim T[\mathbf{n}]) = T[\mathbf{n}] \tag{8.6}$$

where n is the numeral representing n. Along the lines indicated above one can see that $\sim T[\mathbf{n}]$ is true but that the sentence

$$T(g(\sim T[\mathbf{n}])) = T[\mathbf{n}] \tag{8.7}$$

which attributes the truth predicate to its Gödel number is false.

Thus the game-theoretical truth predicate I am considering cannot do its job for all set-theoretical sentences. It will fail to yield the right truth-condition for the liar-type sentence S_n whose Gödel number is the number n yielded by the diagonal lemma. In this sense, the model theory in question is inevitably incomplete. This destroys the main rationale of set theory, namely, its role as the allegedly canonical framework for model theory. The failure of any self-applied set-theoretical truth predicate means that set theory cannot adequately serve as its own model theory, any more than ordinary first-order logic can.

But incompleteness is the least of the woes of axiomatic set theory. The failure of Tarski's T-schema for the sentence S (constructed by means of the diagonal lemma) means that there inevitably exists in any model of set theory a set-theoretical sentence which is true in the ordinary sense of the word but whose truth-condition, when expressed set-theoretically, is false. Now the truth-condition of only first-order sentence asserts the existence of the Skolem functions that codify a winning strategy for the initial verifier in the semantical game connected with it. Considered as a set-theoretical statement, S can fail to be true only if the truth-making strategy functions (Skolem functions) do not exist in a model of the set theory in question, even though they are extensionally possible "functions in extension", as Russell would have expressed the point. And such a failure means that in axiomatic set theory, a sentence is not always equivalent to the set-theoretical counterpart of its second-order translation.

This observation has several different consequences illustrating the inadequacy of axiomatic set theory. What just has been seen means that intuitively speaking my neo-Gödelian sentence S_n should be true, in that the Skolem functions which it asserts to exist in set theory actually do exist in a purely extensional sense. And axiomatic set theory was originally calculated to capture precisely all extensionally possible sets – or at least as many of them as we can.

This might not yet seem to set the Skolem functions asserted to exist in S_n apart from the nonexisting sets and functions which are ruled out by the well-known paradoxes. But the present failure is far more poignant than the traditional ones. What is so special about the failure of the equivalence of S_n and its truth-condition T[n]? One thing that is special about them is that equivalences between a first-order sentence and its second-order truth-condition were seen earlier in Chapter 2 to be nothing but instances of the axiom of choice or of its generalizations. These generalizations can be justified in terms of my game-theoretical approach in the same way as the axiom of choice itself. In other words, in equivalences of sentences like my S_n with their truth-conditions we are merely dealing with instances of the translations of generalizations of the axiom of choice into a set-theoretical language. And yet some of them were seen to fail inevitably. Hence the failure of these generalizations in axiomatic set theory shows that the axiom of choice should be really rejected by axiomatic set theorists, for its motivation is, in effect, rejected by them. To put the same point in other terms, the fact that the axiom of choice is compatible with the usual axioms of set theory (see Gödel 1940) is merely a lucky accident having nothing to do with the theoretical motivation of the axiom of choice. One can in fact say that the spirit of axiomatic set theory is deeply antithetical to the axiom of choice. Since this axiom was seen to have an impeccable theoretical motivation, we have here a strong reason to be wary of axiomatic set theory in general.

As a by-product of this line of thought, we have arrived at an argument to show that my generalization of the axiom of choice is actually stronger than the axiom itself. This verifies the surmise mentioned in Chapter 2. For the axiom of choice, even in its global form, holds at least some of the models of axiomatic set theory, whereas some instance or other of its generalization has been seen to fail in each model.

This point can be put in a different way. Gödel advocated the cultivation of our "mathematical intuition" in such a manner that it will hopefully show us the validity of new axioms of set theory. One of the most firmly entrenched sets of intuitions we seem to have in set theory are the ones that lead to the axiom of choice. I have shown above in a Gödelian spirit that these same intuitions justify stronger

set-theoretical assumptions as well, namely, the set-theoretical counterparts to equivalences between a first-order set-theoretical sentence
and its second-order truth-condition. However, these stronger set-
theoretical assumptions turn out to contradict the received axioms
of set theory. This means that a most natural implementation of
Gödel's idea does not lead to an improved set theory, but to the
rejection of first-order axiomatic set theory.

Here the reader can perhaps begin to see the general reasons for
my suspicions of axiomatic set theory. It was originally constructed
to embody precisely the way of thinking represented among other
things by the axiom of choice. As it flexes its deductive muscle, it
clearly begins to turn menacingly against the intentions of its origi-
nators. It is turning into a veritable Frankenstein's (Fraenkelstein's)
monster which threatens to destroy what it was constructed to serve
and to defend.

Thus we have arrived at a serious indictment against axiomatic set
theory. For each model of axiomatic set theory, one can find a
sentence S in the language of a set theory which is true interpreta-
tionally but false in that model. In brief, axiomatic set theory cannot
be true on its own intended interpretation. When we try to construct
the liar paradox for axiomatic set theory, the liar turns out to be
axiomatic set theory itself.

A possible rejoinder to this indictment is to say: What else is
new? The old paradoxes already show that there are set-theoretical
propositions which we are intuitively tempted to accept but which
cannot be assumed to be true in the models of set theory. This
rejoinder does not cut much ice, however. In the old paradoxes, we
are typically dealing with the failure of an instance of the unrestricted
axiom of comprehension to be true. In other words, we are dealing
with an apparently well-formed concept whose extension is not
well-defined. For instance, there is no well-defined class which is an
extension of $\sim(x \in x)$. In contrast the extensions of the functions
whose existence is asserted by the sentence S conducted above are
unproblematic. The only problem is that these classes cannot be
reified into sets which could exist in the models of the set theory in
question.

More generally speaking, the earlier paradoxes of set theory are
all caused by there being, so to speak, too many sets required to exist
by some set-theoretical assumption or the other. The charge I am
making against the usual formulations of set theory is that it does not

allow the existence of functions (sets) which ought to exist, and in this sense assumes too few sets to exist.

Indeed, what leads one to say that the Skolem functions that are asserted to exist in S actually do so are not a Gödelian set theorist's mysterious intuitions. They exist by the same criteria as are applied every day by mathematicians whenever they speak of the existence of sets and functions. (At least, all classical mathematicians use such criteria; and I will take care of the constructivists and the intuitionists in the last two chapters of this book.) And the criteria they explicitly or (more likely) implicitly use do not in the present case depend on any assumptions concerning the existence of sets or of any other higher-order entities. This can be seen by observing that the second-order sentence $S^{(2)}$ which asserts the existence of the relevant Skolem functions for $\sim T[\mathbf{n}]$ is naturally of the Σ_1^1 form. Hence it is logically equivalent to an IF first-order sentence F. Since F is first-order, its truth is a purely combinatorial matter, independent of what one thinks of the existence or nonexistence of sets and functions.

Thus the new difficulties I have pointed out in axiomatic set theory are not merely an extension of the well-known traditional ones. The only reason why they do not make axiomatic set theory totally useless for all mathematical purposes in one fell swoop is that they concern the entire set-theoretical universe. This is because a self-sufficient truth definition must involve an entire model. But what if you try to formulate a truth predicate for those parts of the set-theoretical universe that are supposed to be used directly in mathematics, such as the theories of natural numbers, real numbers, real functions and so forth? The answer is that such a truth predicate must refer to other parts of the given model of set theory. For instance, for elementary number theory, the Skolem functions over which you are quantifying in your truth predicate are second-order entities, not natural numbers. Hence axiomatic set theory escapes the ultimate ignominy of implying false propositions about ordinary mathematical entities only by a tacit stratification of one's mathematical universe into what, in effect, corresponds to a logician's types. But such a stratification itself militates against the allegedly type-free spirit of set theory, which was supposed to be type-free.

The reaction of a confirmed set theorist to such criticisms of his or her métier might perhaps be an instance of the old rule: If you cannot beat them, join them. In the next chapter it will be shown that IF first-order logic greatly extends the sphere of what logic can do in the foundations of mathematics. Why not join forces and simply harness IF logic to the service of axiomatic set theory? Alas, this strategy fails unless the foundations of axiomatic set theory itself are radically revised. The rest of this chapter is devoted to exposing the difficulties of any straightforward merger of IF logic and set theory.

There is in fact an important reason why axiomatic set theory in its present form is incapable of using the help offered by IF logic. This handicap is due to certain basic features of axiomatic set theory. As was pointed out earlier in this chapter, the main strategy of set theorists is to capture sets as fully as possible as extensions of formulas of the language of set theory. The use of the axiom of comprehension noted earlier is but one manifestation of this strategy. The fundamental fact nevertheless is that this technique becomes largely inapplicable when IF first-order logic is being used. If brief, the usual techniques of axiomatic set theory are in a sense incompatible with IF first-order logic.

What I mean by this is the following: When a set α is captured by means of an ordinary first-order formula $F[x]$, this can be expressed by the general equivalence

$$(\forall x)(x \in \alpha \leftrightarrow F[x]) \tag{8.8}$$

This equivalence can be used as a premise in set-theoretical reasoning. But if $F[x]$ is an irreducible independence-friendly formula, then (8.8) fails to be true. The reason is that it implies that $F[x]$ obeys the law of excluded middle for all values of x, which an irreducibly independence-friendly formula does not do, as was explained in Chapter 7.

The underlying reason for this situation is that in the usual axiomatic set theories, the membership relation $\in$ is construed as a primitive predicate. Every assertion of set membership is therefore an atomic formula. And the law of excluded middle applies to atomic formulas. Hence we can never truly say in a set-theoretical language that a set captures the extension of an irreducibly independence-friendly formula. Hence the entire strategy of trying to capture sets as extensions of formulas fails to do its job when the logic it is based on includes independent quantifiers.

This fact is but a particular manifestation of a more general phenomenon. One of the most crucial questions in the foundations of axiomatic set theory is how sets are to be captured. As was indicated in the beginning of this chapter, the most common type of answer we find in the literature is: As extensions of predicates well-defined in the language of set theory, or at least as extensions of some suitably restricted well-defined predicates. Perhaps the clearest expression is Skolem's and Weyl's explication of Zermelo's (1908) notion of a *definit* predicate. Zermelo had introduced this notion to capture the sets whose existence is asserted by his *Aussonderungsaxiom*. Skolem (1922) and Weyl (1917) in effect identified *definit* predicates with those representable in the language of their first-order set theory.

More generally, there is an influential tradition which equates the existence of higher-order entities with their representability in a suitable language. This tradition has influenced in various ways the development of axiomatic set theory. It is nevertheless hard to find theoretically motivated reasons for the theses of this tradition.

Now we are beginning to see why. When we move from ordinary first-order logic to IF first-order logic, every reason disappears for thinking of well-formedness of complex predicates as the benchmark of their capability to capture a set. Assume that $S[x]$ is a complex predicate of an IF first-order language, with x as its only free individual variable. Then the corresponding set s presumably would be characterized by the sentence

$$(\forall x)(x \in s \leftrightarrow S[x]) \tag{8.9}$$

But this sentence can be true only if $S[x]$ satisfies the law of excluded middle, which is often not the case in IF first-order languages. Hence the extension of $S[x]$ is a set only if $S[x]$ obeys *tertium non datur*. Even if $S[x]$ is perfectly well-formed, it does not capture a set unless it is a formula of an ordinary first-order language. Thus in an IF first-order language of set theory, the well-formedness of a formula does not guarantee the existence of the corresponding set.

This failure of the purported link between well-formedness and set existence does not have anything to do with the reification involved in the idea of sets as individuals over which one can quantify. Suppose that we modify (8.9) so that, instead of characterizing the set s, it now defines a new one-place predicate P, for instance as follows:

$$(\forall x)(P(x) \leftrightarrow S[x]) \tag{8.10}$$

Then we are not doing any better than with (8.9). For, like (8.9), the explicit definition (8.10) presupposes that $S[x]$ conforms to the law of excluded middle.

The trouble here cannot be blamed on the absence of a purely truth-functional equivalence, either. In order to see this, we may introduce a new connective $\Leftrightarrow$ defined semantically as follows:

$S_1 \Leftrightarrow S_2$ is true if and only if   (8.11)

(a) both S_1 and S_2 are true

or

(b) both S_1 and S_2 are not true (i.e., either false or neither true-nor-false)

What will happen? if we now use $\Leftrightarrow$ instead of $\Leftrightarrow$ in the explicit definition of a new predicate? Consider the result of this change in (8.10), that is

$$(\forall x)(P(x) \Leftrightarrow S[x])$$ (8.12)

Intuitively speaking, for each value b $P(b)$ is true if and only if $S[b]$ is true; otherwise it is false. Let us now apply the definition schema (8.12) to the truth predicate $Tr[x]$. Then there is in the extended language a predicate constant $P(x)$ such that

$$(\forall x)(P(x) \Leftrightarrow Tr[x]).$$ (8.13)

By the diagonal lemma, there is a number n such that n is the Gödel number of $\sim p(n)$, where n is the numeral representing n. If $\sim P(n)$ were true, it would follow from (8.13) that it is false. But if $\sim P(n)$ were not true, it would likewise follow from (8.13) that it would be true. Hence definitions of the form (8.12) are not always admissible.

The existence (admissibility) of definitions like (8.12) is what Russell's Axiom of Reducibility (Russell and Whitehead 1910–13, Vol. 1, pp. 55–59) is supposed to guarantee. What has been found is that in IF first-order logic no principle remotely like Russell's reducibility axiom holds.

In different terminology, we can say that the failure of well-formedness to guarantee set or predicate existence is no abject that even the safest vehicle of well-definedness, an explicit definition, does not guarantee set existence or even the existence of a quantifier-free predicate.

These observations are most suggestive. They show that some of the favorite methods used by logicians and mathematicians cannot be extended beyond ordinary first-order theories. Explicit definitions cannot be used without strict precautions outside ordinary first-order logic. And the tactic of set theorists in trying to access sets via the predicates whose extensions they are supposed to be does not always work after the possibility of informational independence is recognized. This in turn shows that Frege's entire approach to the foundations of logic and mathematics is subject to severe limitations, for the only role that sets play in his approach is as course-of-values of predicates ("functions").

Another way of putting essentially the same point is to say that the usual theory of definition is one of the many features of ordinary first-order logic that cannot be extended beyond its idiosyncratic boundaries.

There is a more upbeat way of looking at these results. What has been found is that the power of IF first-order logic in dealing with complex predicates extends way beyond predicates that are well-defined for the purposes of set theorists. Hence once again IF first-order logic has been found to be a better conceptual tool in the foundations of mathematics than received set theory.

This suggests a major new way of improving the existing techniques of set theoretical conceptualization and argumentation. This way consists in departing radically from Cantor's requirement that set membership must be well-defined, and in allowing for sets which are only partially defined in the sense that the law of excluded middle does not apply to membership in such a set: besides definite members and nonmembers there are also potential elements which are neither. Such a set theory would be in the spirit of the recently kindled interest among logicians, mathematicians and formal linguists in partiality, evinced and surveyed by Fenstad (1996). Indeed, such an IF set theory would provide a convincing example of the ways in which allowing partiality can actually strengthen significantly one's conceptual resources.

A few further comments may be in order. Set theory is sometimes accused of reifying sets into objects. It is not clear precisely what such accusations can mean. The treatment of membership relation as a primitive relation on a par with others offers a possible precisation of such accusations. For what it amounts to is a treatment of the relation of a set to each one of its members in the same way as any relation between two individuals (objects) is treated. And if this is what the reification of sets means, it is indeed a mistaken or at least ill-advised procedure, for it cuts axiomatic set theory off from some of its most important potential resources.

Axiomatic set theory is thus subject to various theoretical problems as a framework for doing mathematics, and may even turn out to be totally unacceptable in its present form. Furthermore, I have not as much as touched on the difficulties of actually choosing the axioms of set theory. What assumptions about set existence should we make? How can we decide? Enough has been said in any case to justify a search for a better framework for mathematical theorizing.

Note

[1] This conception of mathematical existence as pertaining to some kind of super-universe of all possible structures played an important role in the foundations of mathematics around the turn of the century. It was seldom, if ever, discussed explicitly, and later philosophers and historians have neglected it virtually completely. Yet it rears its head in the writings of the foremost mathematicians and logicians discussing the foundations of mathematics of that period. The following is a brief list of some of the more conspicuous appearances of this idea: Dauben (1979, pp. 145–146, on Cantor); Frege and Hilbert (Frege 1980, pp. 43–44); Poincaré (1905–6, p. 819; 1952, pp. 151–152); Tarski (1935, p. 318; 1956, p. 199).

9

IF Logic as a Framework for Mathematical Theorizing

In earlier chapters, especially in Chapters 3, 4, and 7, IF first-order logic was introduced and its mathematical properties were briefly studied. The crucially important question here is, nevertheless, not what mathematics can do for your logic, but what your logic can do for mathematics. What can logic – any logic – in principle do in and for mathematics?

Here we are in the proximity of some of the main traditional, not to say nostalgic, issues in the foundations of mathematics. For one thing, the logicist philosophers used to have a simple answer to the question just posed. Their answer was: Everything. Or, as it also used to be put: Mathematics can be reduced to logic. Unfortunately, the nostalgia of this answer is not what it was in the good old days. It was then thought that "the logic" could be formulated as an axiom system on the same level as different mathematical axiom systems, so that the purported reduction could be discussed as a reduction of one axiomatic system to another.

What has been found in this book shows the hopelessness of such thinking. Not only is a complete axiomatization impossible for the basic part of logic, IF first-order logic but even when there exists an axiomatization of some part of logic, it is an axiomatization in a radically different sense from axiom systems for nonlogical theories. This disparity makes it nonsensical to speak of a reduction of a nonlogical axiom system to a logical one.

Yet the issues raised by the logicists are relevant and important. However, instead of trying to formulate them as questions of the reducibility of one axiom system to another, we should formulate them as questions about the status of mathematical concepts and mathematical modes of reasoning vis-à-vis logic. Thus we may – and must – ask two different questions:

(a) Can one define the crucial mathematical concepts in logical terms?

(b) Can one express the modes of inference used in mathematics in logical terms?

It is important to realize that what has been found in this book changes the received force of the questions (a)–(b). In particular:

(i) In reality, (a) involves two different questions. They are: (a1) Can one define in purely logical terms such concepts as natural number and real number? (a2) Can one formulate a descriptively complete axiom system for different mathematical theories?

(ii) Since logic itself is incomplete, the force of (b) is not to ask whether mathematical modes of inference can be captured by deductive rules of logic, but rather whether they can be captured as semantically valid logical inferences.

Here we encounter a problem situation that has prevailed in twentieth-century philosophical thinking since the demise of the older logicist program. A philosopher's typical idea of logic is the ordinary first-order logic. There is in fact much to be said for it as a framework for mathematical theorizing. Because this logic is semantically complete, a descriptively complete axiomatization of any given mathematical theory by means of ordinary first-order logic automatically yields a deductively complete axiomatization. Most philosophers would in fact agree that it would be splendid if we could do all of mathematics within the framework of first-order logic – *wenn das Wenn im Wege nicht wäre*. But the sad truth is that all of mathematics just cannot be done in terms of ordinary first-order logic, or apparently on the first-order level in general. There are crucially important concepts and modes of inference in mathematics that cannot be captured by means of ordinary first-order logic. They include the principle of mathematical induction and the notions of finiteness, infinity, well-ordering, cardinality, power set, and so forth. And the way these notions are not captured by ordinary first-order logic is much deeper than merely not being able to set up complete axiom systems incorporating these notions. The impossibility is model-theoretical. By means of ordinary first-order sentences one just cannot capture the right class of structures as their models. For one uncharacteristically simple example, there is no formula of ordinary first-order logic with identity = as its only predicate which is true if and only if the universe of discourse is infinite.

Jon Barwise has expressed this shortcoming of ordinary first-order logic as follows:

Paging through any modern mathematical book, one comes across concept after concept that cannot be expressed in [ordinary] first-order logic, concepts from set-theory (like *infinite set, countable set*), from analysis (like *set of measure 0 or having the Baire property*), and from probability theory (like *random variable* and *having probability greater than some real number r*) are central notions in mathematics which, on the mathematician-in-the-street view, have their own logic. (Barwise in Barwise and Feferman 1985, pp. 5–6)

I will return to Barwise's statement later.

The need to cope with all these different kinds of mathematical concepts and mathematical modes of reasoning imposes requirements on logic that ordinary first-order logic cannot satisfy. But it is not clear what kind of logic, if any, or what kind of use of logic can do so. This is what I meant by the current tension between mathematics and logic in the foundations of mathematics.

Because of this failure of ordinary first-order logic to serve the needs of mathematics, it is usually thought that mathematical reasoning is irreducibly set-theoretical or that it involves otherwise higher-order notions like concepts, functions or predicates. This lends a tremendous prima facie importance to the third use of logic in mathematics mentioned at the end of Chapter 1, namely, to the use of logic (usually ordinary first-order logic) as a framework for doing axiomatic set theory. Indeed, we have here an explanation for the allegedly central role of set theory in the foundations of mathematics. Even though there is little agreement among the philosophical theories, in practice there is a great deal of uniformity as to how mathematics should in principle be done. It is precisely axiomatic set theory which is generally (but not universally) thought of as the framework in which all serious mathematical theorizing should – or at least could – take place. Unfortunately, it was found in the preceding chapter that axiomatic set theory in its present form is not capable of doing what it must be capable of doing to qualify as a framework for mathematical theorizing. Hence we have to look for some other sources of greater logical horsepower than what is provided for by ordinary first-order logic.

How soon the need for a standard higher-order logic arises can be seen by considering a Frege-type definition of number as the class of all equicardinal sets. Equicardinality cannot be defined in ordinary first-order terms. We not only use higher-order logic in the customary definition of equicardinality which says that two classes are equicardinal if and only if there exists a one-to-one relation mapping them onto each other. The notion of higher-order existence must be taken in the standard sense. For otherwise there might very well exist two classes which are equicardinal in the intended pretheoretical sense but which do not have any nonstandardly existing relation to map them on each other one-to-one. (In the most common variant of nonstandard higher-order logic, the higher-order variables range over those higher-order entities that can be defined in the theory in question.) If brief, equicardinality is apparently an essentially (that is, standardly) higher-order or perhaps set-theoretical notion. And if so, Frege's and Russell's famous reduction could not reduce mathematics all the way to logic (even if it were otherwise successful), only to the branch of mathematics called set theory – or misleadingly labeled "higher order logic". Thus no reduction of mathematics to anything else is accomplished – or so it seems.

Here IF first-order logic puts things in a radically new light. To take but one example, even though it may be a short step for one logic, it is a potentially long step for mathkind to discover that equicardinality can be defined in IF *first-order* logic for the extensions of two ordinary first-order predicates, simple or complex, say $F_1[x]$ and $F_2[x]$. The following formula will serve the purpose:

$$(\forall x)(\forall z)(\exists y/\forall z)(\exists u/\forall x)((F_1[x] \supset F_2[y]) \;\& $$
$$(F_2[z] \supset F_1[u]) \;\& \;((y = z) \leftrightarrow (u = x))) \tag{9.1}$$

A moment's thought will convince you that (9.1) does the job it was supposed to do. This thought might be facilitated by the second-order translation of (9.1):

$$(\exists f)(\exists g)(\forall x)(z)(F_1[x] \supset F_2[f(x)] \;\& $$
$$(F_2[z] \supset F_1[g(z)]) \;\& \;((f(x) = z) \leftrightarrow (g(z) = x))) \tag{9.2}$$

As a special case we obtain an expression for the equicardinality of two sets α and β.

$$(\forall x)(\forall z)(\exists y/\forall z)(\exists u/\forall x)((x\in\alpha \supset y\in\beta) \,\&$$
$$(z\in\beta \supset u\in\alpha) \,\& \,((y=z)\leftrightarrow(u=x))) \tag{9.3}$$

If we try to remove the restriction of $F_1[x]$ and $F_2[x]$ to ordinary first-order predicates, then we will find out that our definition works only if $F_1[x]$ and $F_2[x]$ are ordinary first-order formulas.

One interesting fact here is that the equicardinality of two sets α and β cannot be expressed by an ordinary first-order formula even for the cases where α and β are finite. Hence we have here an example of the ways in which IF first-order logic gives us additional power even for the study of finite structures.

Similar things can be said of the notion of infinity. In mathematics, we are dealing with an infinite universe. But even before we can face the question as to whether or not the infinity of the universe of mathematics has to be assumed or the question whether this is a logical or a mathematical assumption, we have to see whether or not the assumption itself can be expressed in purely logical terms, or whether the mere formulation of the concept of infinity requires the help of set theory. In ordinary first-order logic, the infinity of the universe cannot be expressed without the help of nonlogical constants. However, in IF first-order logic the infinity of the universe of discourse can be expressed by using $=$ as the only predicate. The following is a case in point:

$$(\forall x)(\forall z)(\exists y/\forall z)(\exists u/\forall x)((x \neq y) \,\& \,(z \neq y) \,\& \,((x = z)\leftrightarrow(y = u))) \tag{9.4}$$

This is easily seen to be equivalent to

$$(\exists f)(\forall x)(\forall z)((x \neq f(x)) \,\& \,((f(x) = f(z))\leftrightarrow(x = z))) \tag{9.5}$$

If (9.4) or (9.5) is true, then the universe must be either empty or infinite.

By relativizing the quantifiers in (9.4) to the extension of a given ordinary first-order predicate $F[x]$ (simple or complex) we can likewise define the infinity of this extension in first-order terms. If the restriction to ordinary first-order formulas is removed, we obtain a characterization of the infinity of the class of individuals which does not make $F[x]$ false.

It is of some philosophical interest to take a closer look at sentences like (9.4). This sentence contains no predicates other than the identity relation. Yet it has a highly nontrivial meaning. Where does this nontrivial meaning come from? The only candidates are the notion of identity and quantifiers (plus propositional connectives). Their meaning thus has to be understood independently of any constant predicates. This suggests an interesting conclusion concerning the notion of identity which is codified in first-order logic. This notion is not based on, or otherwise derivative from, any particular constant predicates, such as so-called sortal predicates. The notion of identity in a given domain can be understood independently of one's understanding of any particular predicates of the members of that domain – indeed, independently of one's understanding of any nonlogical predicates at all.

Moreover, a wealth of other mathematical concepts can be defined in IF first-order logic or in its extended version. They include the following:

(i) The notion of well-ordering. Consider a discrete ordering defined by a (possibly complex) relation $R[x, y]$ expressible in ordinary first-order language. We can then express in an IF first-order language the fact that $R[x, y]$ is *not* a well-ordering. For the purpose, what we have to express is that there exists an infinite descending sequence of individuals. This can be expressed as follows:

$$(\forall x)(\forall z)(\exists y/\forall z)(\exists u/\forall x)(x \neq y \ \& \ z \neq u \ \& \ R[y, x] \ \&$$
$$R[u, z] \ \& \ (x = z \leftrightarrow y = u)) \tag{9.6}$$

A moment's reflection easily shows that (9.6) can be used to express the failure of R to be a well-ordering, in that it says that there is an infinite descending sequence of individuals in the order R. Hence the concept of well-ordering can be expressed in extended IF first-order logic.

(ii) Likewise, the principle of mathematical induction can be formulated in extended IF first-order logic, while its contradictory negation is expressible in (unextended) IF first-order language.

(iii) The notion of power set can be characterized in extended IF first-order logic in a sense (see here Krynicki and Väänänen, 1989). The sense in question is that for two simple predicates $A(x)$, $B(x)$ we can express by means of extended IF first-order logic that the cardinality of B(x) is 2^α, where α is the cardinality of A(x).

This result shows incidentally that the Löwenheim–Skolem theorem does not hold in extended IF logic.

(iv) The Bolzano–Weierstrass Theorem is expressible in extended IF first-order logic. This theorem says that every bounded infinite set of real numbers has an accumulation point. It has accordingly the form

$$(\forall X)((\text{Inf}(X)\,\&\,\text{Bound}(X))$$

$$\supset (\exists y)(y \text{ is an accumulation point for } X)) \qquad (9.7)$$

Here the notion of boundedness (Bound) and accumulation point can be defined in ordinary first-order logic. In (i) above, it was seen that infinity can be defined in (unextended) IF first-order logic, ergo by a Σ_1^1 formula. Hence the Bolzano – Weierstrass Theorem can be formulated as a Π_1^1 formula. This suffices to show that the theorem can be formulated in extended IF first-order logic.

(v) In topology, the notion of open set can be characterized by means of IF first-order logic. For instance, we can say that X is open if there is a function that maps each element of X on a neighborhood which it is a member of and which is a subset of X. This can be expressed by a Σ_1^1 second-order formula, ergo by an IF first-order formula.

(vi) In the same direction, the topological notion of continuity can be characterized in extended IF first-order logic. In order to see how this can be done, it suffices to recall that a function f is continuous in the topological sense if and only if the inverse of each open set is open. By what was found in (v), this is expressible in truth-functionally extended IF first-order logic.

(vii) A number of apparently mathematical (set-theoretical) truths also turn out to be truths of IF first-order logic. A case in point is the principle of transfinite induction. It can be taken to be an inference from

$$(\forall x)(\forall z)(\exists y/\forall z)(\exists u/\forall x)[(x = z \leftrightarrow y = u)\,\&\,(A(x) \supset A(y))\,\&$$

$$(A(z) \supset A(u))\,\&\,y < x\,\&\,u < z] \qquad (9.8)$$

to

$$(\forall x)(\forall z)(\exists y/\forall z)(\exists u/\forall x)[(x = z \leftrightarrow y = u)\,\&\,y < x\,\&\,u < z] \qquad (9.9)$$

In (9.8)–(9.9) the ordering relation is taken to be $<$. An explanation might run as follows: The contradictory negation of (9.9) says that there is no infinite descending chain. The contradictory negation of (9.8) says that there is no infinite descending chain of As – in other words, that there is a smallest A.

These examples show that much more mathematics can be formulated by means of extended (and in some cases even unextended) IF logic than can be formulated in ordinary first-order logic. It is amusing to compare the above sample of examples (i)–(vii) with Jon Barwise's list (quoted earlier in this chapter) of notions that transcend the powers of logic in its accustomed sense. Several of his prize specimens of characteristically mathematical concepts have been shown to be expressible in extended or unextended IF first-order logic. In Chapters 3 and 4, it was shown that these logics are so-called with an as good or better right than ordinary first-order logic.

In conventional philosophical terms, the tremendous increase in conceptualizing power which IF first-order logic gives could perhaps be said to aid and abet a logicist philosophy of mathematics. Such statements have to be taken with a pinch of salt, in that the historical movement known as logicism was using higher-order logic (and calling it "logic," instead of restricting the term to first-order conceptualizations) and was also committed deeply to a nonstandard interpretation of their higher-order logic. However, in the reconstructed sense of logicism explained earlier in this chapter, the increased power of IF first-order logic as compared to ordinary first-order logic shows that the new logic indeed helps greatly the logicist cause. Both conceptualizations and modes of inference which were earlier impossible to capture on the first-order level can now be captured by means of IF first-order logic.

What has been found in any case forces us to radically revise our ideas of the borderline between logic and mathematics. Several of the most crucial concepts and modes of inference of mathematics that have been believed to be beyond the powers of logic to capture have turned out to be expressible in IF first-order logic.

For instance, the main obstacle to a purely logical definition of number has been the impossibility of characterizing equicardinality in terms of ordinary first-order logic. It is generally thought that such a characterization inevitably involves quantification over higher-order entities. IF first-order logic provides a way of defining equicardinality in entirely logical terms and hence removes this particular obstacle to a purely logical definition of number.

Similarly, the observations made above concerning the principle of mathematical induction have sometimes been claimed to be a specifically mathematical principle of reasoning. Now it can be seen that this principle can be formulated on a first-order and therefore uncontroversially logical level.

Likewise the notion of power set is one of the most crucial in set theory. The possibility of characterizing it and the other notions dealt with above, may thus be seen as a partial and qualified vindication of the belief of early set theorists that they were dealing with purely logical conceptualizations.

Hence examples like (ii)–(vi) are not just isolated instances of what can be done by means of IF first-order logic. They go a long way toward showing that the inevitable shortcomings of axiomatic first-order set theory can be overcome by means of IF first-order logic. For instance, the existence of uncountable sets can be guaranteed by means of suitable axioms that can be formulated in terms of extended IF first-order logic. This follows from what was said earlier about the power set (see (vi)). My point is not spoiled by the fact that the existence of nondenumerable sets cannot be expressed in an unextended IF first-order logic, since such a logic admits of the Löwenheim–Skolem theorem. Notwithstanding this restriction, the extended IF first-order logic was seen to provide the requisite means of capturing such notions as well-ordering.

Thus IF first-order logic apparently can help set theory precisely where it needs help most badly, namely, in providing standard formulations of central mathematical concepts. This suggests that the next step forward in the foundations of mathematics might very well be to simply combine axiomatic set theory and IF logic or, rather, to use IF first-order logic as the basic logic of our familiar axiomatic set theory. This step looks tempting but it is not feasible, as was shown in the preceding chapter.

In other respects, too, it might very well look as if the euphoria that was generated by the first victories of IF first-order logic is soon dissipated. Not only is it impossible to combine ordinary axiomatic set theory with this new logic. What is more, IF first-order logic is equivalent only to a small fragment of second-order logic, namely, the Σ_1^1 fragment. Even by means of extended IF first-order logic, our logic can only capture the $\Sigma_1^1 \cup \Pi_1^1$ fragment of second-order logic.

Indeed, second-order logic (and, more generally speaking, higher-order logic) seems to be emerging as the superior rival to IF first-order logic. In the preceding chapter, axiomatic set theory in its present form was found to be seriously wanting as a framework for working mathematics. But the criticisms offered there do not apply against second-order or higher-order logic. On the contrary, what has been found in this book might first seem to enhance the status of second-order logic in the foundations of mathematics. For instance, in Chapter 5 I emphasized the importance of putting forward descriptively complete (albeit deductively incomplete) theories to guide us in extending the scope of actual deductively developed mathematics. In the preceding chapter, it was shown that axiomatic set theory cannot perform this service in its present form. A much more promising candidate for this job is offered by second-order logic, of course with what Henkin (1950) called the standard interpretation. (This interpretation was explained briefly in the preceding chapter.) Stewart Shapiro has recently (1991) argued ably for second-order logic as the appropriate "foundations without foundationalism" for mathematics. Indeed, a great deal of mathematics can be formulated in second-order terms in a much stronger sense than set theory allows us to do. Not only does second-order logic provide you with languages in which practically any mathematical theory can be expressed – what is much more, this second-order formulation, unlike the set-theoretical one, is model-theoretically faithful to the original. In other words, when a mathematical statement S can be formulated in second-order logic it normally has precisely the models it is normally thought of as having. Hence, for a mathematical statement S to be a theorem of a theory characterizable by a conjunction of axioms T all of which are expressible in second-order language, it is necessary and sufficient that the conditional $(T \supset S)$ be logically true. All questions concerning such a mathematical theory can thus in principle be reduced to questions concerning the logical truths of second-order logic.

As was mentioned earlier, this remark seems to apply to practically all mathematical theories and hence to mathematical propositions that can be formulated in them. Cases in point include, for instance, many famous unsolved mathematical propositions from

Goldbach's Conjecture to the Special Continuum Hypothesis. In principle, though not necessarily in practice, the second-order logical formulation provides a clear-cut goal for a search for a proof of such conjectures. Needless to say, the guidelines we can hope to obtain are indirect ones. They are criteria for the acceptability (validity) of new rules of inference that can facilitate actual deductive proofs, rather than tactical principles for carrying out proofs by means of already known and accepted rules.

Just how much mathematics can thus be handled in second-order logic requires careful examination. It is not difficult to argue, as is for instance shown in Shapiro (1991, Ch. 5; cf. also Shapiro 1985) that all the most common set-theoretical conceptualizations, including minimal closure, cardinality, continuum hypothesis, well-ordering, axiom of choice, and well-foundedness, can all be captured by means of second-order logic. By their means, obviously a great deal of actual mathematics can be formulated. It looks pretty much as if the entire classical analysis can be formulated in second-order terms. It is also easy to see that general topology can be done in this framework. I will return to the limitations of such formulations at the end of this chapter.

But many hard-nosed logicians will not be happy with the proposal of using a second-order language as a medium for their mathematical theorizing, and for a good reason. In order for such a language to serve its purpose, its second-order variables must be taken in their standard sense. They must be taken to range over *all* extensionally possible entities of the appropriate type (sets, functions, etc.). For instance, one's function variables have to be taken to range over all arbitrary functions, however noncomputable (cf. here Hintikka 1995a). But if so, we face all the problems connected with the ideas of arbitrary set and arbitrary function (strictly speaking all arbitrary subsets of a given infinite set and all arbitrary functions from a given infinite set to another given set). I can indicate this kind of commitment to arbitrary higher-order entities by saying that it involves the idea of "all sets". Another way of expressing myself might be to speak of the standard interpretation in Henkin's sense. But whatever the name that this idea passes under, its smell is equally foul to many logicians. And there is a great deal to be said for their perceptions. The idea of the totality of all (sub)sets is indeed a hard one to master. Therefore, even though a second-order formulation of mathematical propositions and conjectures may be superior to treating them in an axiomatic set theory, it involves further serious problems.

Higher-order logic, like axiomatic set theory, can of course be given a nonstandard interpretation. Indeed, the best known incarnation of higher-order logic, the theory of types of Russell and Whitehead (1910–13), was intended to have a nonstandard interpretation (cf. Ramsey 1925).[1] But as Frank Ramsey was not slow to point out, this interpretation is just not in agreement with the mathematicians' prevalent way of thinking. For this reason, the nonstandard interpretation of higher-order logic loses all the advantages it lots had on the standard interpretation in guiding mathematicians in their search for stronger axioms of set existence, and for other mathematical and logical truths that could provide stronger deductive principles.

Since nonstandard interpretations do not help us to deal with the problems of set existence, what can? Here IF logic seems to offer its services to us. As long as we can stay on the level of first-order logic, independence-friendly or not, then problems of set existence do not arise. We do not have to open the Gordian knot of set existence since it was not tied in the first place.

What can be done by means of IF first-order logic? Relatively little, it might seem. Second-order logic cannot be reduced in any direct sense to IF first-order logic. As was pointed out, only the Σ_1^1 fragment of second-order logic can be translated into IF first-order logic. The fact that contradictory negations of IF first-order logic are translatable to Π_1^1 fragments of second-order logic, and vice versa, does not change the picture essentially.[2]

However, there is an indirect sense in which mathematical theories and mathematical problems can be reduced by means of IF first-order logic. What we can do is to try to reconstruct a given higher-order theory T by means of ordinary first-order logic as a many-sorted first-order theory. In this reconstruction, each of the types needed in T will become a sort of its own. The different types can be thought of as forming a special syncategorematic sort (set) τ. The structure of this set can be expressed by means of explicit first-order axioms. Furthermore, we can introduce a function t which maps each entity (other than a type) on its type. Each sort corresponds to the set of all entities mapped by t on the same type. Among other things, there will then be a defined relation $\sigma(x, y)$ which holds if and only if the type of y is the type of sets of entities of the same type as x. All relations between entities belonging to different sorts can be handled by a suitable new predicate $E(x, y)$ (also written $x \in y$) representing membership. Obviously, we can then rewrite T as a theory T^* about our new many-sorted first-order models. The result will then be somewhat like a stratified set theory, with the different types of sets separated from each other by having different sortal predicates, that is, being mapped by t on different types.

This way of dealing with higher-order logic as if it were a many-sorted first-order logic is in many ways a most natural one. The ease with which we can do it may very well have been one of the reasons why the distinction between first-order logic and higher-order logic crystallized so slowly in the consciousness of most logicians and philosophers, as emphasized by Moore (1988). It is not offered here as any great novelty. It may even be argued that this is not unlike how early contemporary logicians like Frege and Russell viewed their higher-order logics.

In any case, the crucial question is: Why does not the many-sorted first-order theory T^* capture everything there is to be captured by means of the second-order theory T? The answer is that the many-sorted first-order formulation does not implement the standard interpretation which is needed to make T a genuinely second-order theory. For there is in T^* no way of guaranteeing that for each extensionally possible class of first-order entities (individuals or tuples of individuals) there is a second-order entity having them (and only them) as its elements, and likewise for all higher-order entities. But now comes the pleasant surprise. To implement the standard interpretation, it suffices to introduce a finite number of additional second-order axioms, all of which have the form

$$(\forall X)(\forall w)(\exists z)(\exists y)(\forall x)((X(z) \supset (t(z) = w)) \supset (\sigma(w, y) \ \& \\ (x \in y \leftrightarrow X(x)))) \tag{9.10}$$

where $\sigma(x, y)$ is the predicate explained above. If desired, they can even be combined in a single axiom with the same lone initial second-order quantifier $(\forall X)$. But any finite conjunction J of sentences of the form (9.10) is of the Π_1^1 form, and hence a very special kind of second-order sentence.

Here the axioms of the form (9.10) are clearly closely related to the comprehension axiom of axiomatic set theory. However, the comprehension axiom of any first-order axiomatization of set theory (or any substitution-instance of it, if the comprehension principle is formulated as an axiom schema) is an ordinary first-order sentence. Hence it cannot do the same job as (9.10), for the whole point of (9.10) is that the quantifier $(\forall X)$ is a second-order one and it has to be taken in the standard sense.

By conjoining T^* and J we obtain a theory $(J \& T^*)$ which does the same job as the original theory T. It is not only expressible in a second-order language; it is expressible in the Π_1^1 fragment of that language.

In the way – and in the sense – just sketched, the entire higher-order logic can be reduced to the Π_1^1 fragment of second-order logic. The possibility of this "transcendental reduction" (as Husserl might have said here) has not attracted much attention. It nevertheless promises some very real advantages.

The appearance of IF logics on the scene throws new light on the situation. The Π_1^1 character of the reducts means that the new theory is translatable into the language of extended IF first-order logic. In the sense which appears from these observations, every mathematical theory which can be formulated in a higher-order language can also be formulated in extended IF first-order language. This result can be extended to all theories expressible in terms of finite types. As was indicated above, this covers most of the actual mathematical theories. In this qualified sense, virtually all of classical mathematics can in principle be done in extended IF first-order logic.

Furthermore, the question whether a given proposition C follows logically from T is equivalent to the question whether the following sentence is logically true (valid):

$$(J \& T^*) \supset C \tag{9.11}$$

where J is the conjunction of all sentences of the form (9.10) for all the different types (sorts) needed in T^*. Since we are dealing here with normal second-order logic, the conditional can be taken to be the usual truth-functional one, equivalent to

$$\neg(J \& T^*) \vee C \tag{9.12}$$

where $\neg$ is the contradictory negation. Moreover, T^* and C are (many-sorted) first-order sentences. Therefore (9.12) is logically equivalent to a Σ_1^1 sentence. But this means that it has a translation into an (unextended) independence-friendly *first-order* language.

Combined with our earlier remarks on the way mathematical theories and problems can be represented in second-order languages, we thus obtain a remarkable result. In an interesting sense, a great many mathematical *theories* can be formulated in an extended IF first-order language. Such a theory can be represented in the form $\neg\, T$, where T is a sentence of an (unextended) IF first-order language and $\neg$ the contradictory negation. This includes all the mathematics that can be done in terms of a theory of finite types.[3] Likewise, a great many mathematical *problems* can be taken to relate to the logical status of a sentence of an *un*extended IF first-order language. For instance, if the problem is to prove that a conjecture C follows from second-order axioms T, then there is an IF first-order sentence, S such that S is logically true (valid) if and only if C is a logical consequence of T.

The same argument can easily be applied to the entire theory of finite types instead of to second-order logic.

In the special case in which the original mathematical theory T, expressed in the theory of finite types, is categorical, the proposition C is true in the sole model of T (*modulo* isomorphism) iff $T \vdash C$. But this was seen to hold if and only if a certain IF first-order sentence is logically true. Thus for categorical theories mathematical truth can in a sense be equated with logical truth in IF first-order logic. In particular, the truth of any proposition of elementary arithmetic is equivalent to the *logical* truth of an IF first-order sentence. In general, it is an interesting historical question whether this subtle model-theoretical connection between mathematical truth and material truth has tacitly been instrumental in confusing people about the relationship of ordinary truth to logical truth, especially as many of the best known mathematical theories are, when formulated in higher-order terms, categorical. Cases in point are elementary arithmetic, the theory of reals, elementary geometry, and so forth.

For instance, Peano arithmetic becomes categorical when complemented by a second-order induction axiom. This axiom system can be seen to be equivalent to a Π_1^1 sentence, say N_0. Hence the truth of any unproved arithmetical conjecture F, say Goldbach's Conjecture, is equivalent to the validity of $(N_0 \supset F)$. But this is equivalent to a Σ_1^1 sentence and a fortiori equivalent to a sentence F_0 of IF first-order language. The truth of Goldbach's Conjecture is thus equivalent to the validity of an IF first-order formula.

The upshot of this line of thought is thus a kind of reduction of the entire finite theory of types, with standard interpretation, to IF first-order logic. Since most of mathematics can in principle be expressed in a standardly interpreted finite theory of types, this reduction throws some interesting light on mathematics in general. For what can we say of the output sentences of this reduction? They are IF *first-order* sentences. All their bound variables range over individuals. This should warm the heart of every philosophical nominalist. More importantly, their interpretation is completely free of the logical problems that beset the notion of *all subsets* of a given infinite set. An IF first-order sentence is valid if and only if a certain relational structure cannot help being instantiated in every model. The problem whether a given IF first-order sentence is valid or not is therefore a combinatorial problem in a sufficiently wide sense of the term. This sense is in fact not as loose as it might at first be suspected to be. How natural it is to think of what is going on in ordinary first-order logic as being fundamentally combinatorial is shown vividly by Hao Wang's (1990, essays 10–11) reduction of the decision problem for ordinary first-order logic to domino problems. And in this respect IF first-order logic does not differ in principle from the ordinary one. Indeed, it can be shown that the decision problem for IF first-order logic can be reduced to domino problems that incorporate the characteristic failure of perfect information: some specified domino tiles must be fitted in without knowing what has happened elsewhere in the construction (tiling).

Thus, it is not hard to convince oneself that the additional expressive power of IF first-order logic, over and above that of the ordinary one, is combinatorial in nature, rather than set-theoretical. The overt novelty of IF logics is a greater freedom of the various dependence relations between quantifiers and connectives. This is a combinatorial matter, and not a question of the existence of infinite sets of different kinds. IF first-order logic is in fact one of the few extensions of first-order logic which adds to its power in dealing with *finite* structures (models). This additional power is exploited by, among others, Blass and Gurevich (1986). A feeling for this extra force of IF logic can perhaps be garnered also from examples like (D) in Chapter 2.

Conversely, the question whether a sentence of IF first-order logic reduces to an ordinary sentence is a purely combinatorial one. In order to see this, assume that an IF first-order sentence is given in its

Skolem normal form

$$(\forall x_1)(\forall x_2)\cdots(\forall x_n)\cdots(\exists y_i/\forall x_{i1}, \forall x_{i2}, \ldots)\cdots$$
$$S[x_1, x_2, \ldots, x_n, \ldots, y_i, \ldots] \tag{9.13}$$

Here each set $X_i = \{x_{i1}, x_{i2}, \ldots\}$ is a subset of $\{x_1, x_2, \ldots, x_n\}$. A sufficient condition for (9.13) to be equivalent to an ordinary first-order sentence is that the finite sets X_i can be linearly ordered by class-inclusion. It can be shown that this comes close to being the most general sufficient condition. In fact, it is the widest sufficient condition that can be formulated by the sole means of the sequence of prenex quantifiers.

Our reduction is thus in effect a reduction of practically all familiar mathematics to combinatorial theory, in a sufficiently wide sense of the term. This result throws interesting light on the nature of mathematics in general and on its relation to logic. Among other things, it shows that all mathematical reasoning can in principle be considered as being logical in nature. It also shows that the conceptual problems connected with the idea of *all subsets* of a given infinite set can be completely dispensed with in most of the foundations of mathematics. Admittedly, mathematical problems are not automatically solved by reducing them to problems concerning IF first-order logic. However, such a reduction opens various possibilities of conceptualization. For instance, the structure of the reduct (i.e., of the IF first-order sentence which is the output of the reduction process) will reflect the (combinatorial) difficulty of the problem codified in the original sentence.

My approach has affinities with Hilbert's approach to the foundation of mathematics, which I have discussed elsewhere (see Hintikka 1996). Rightly understood, Hilbert's emphasis was not on formalism but on the combinatorial (in contrast to set-theoretical) thinking as the true basis of mathematics.[4] This agrees with my thesis that mathematics is at bottom combinatorial rather than set-theoretical in character. For instance, our truth definition was formulated by reference to the combinatorial properties of a truth predicate as applied to the Gödel numbers of sentences. This is reminiscent of Hilbert's idea of using the combinatorial properties of a completely formal language as the basis of his foundational approach. This idea is not, philosophically speaking, formalistic at all. On the contrary, we have seen that it supports a realistic conception of truth. The main difference between us and Hilbert is that Hilbert thought that *finite* combinatorics is all we need. It is not, and the most interesting aspects of IF first-order logic embody problems of *infinitary* combinatorics. In so far as we can look away from this difference, our approach can nevertheless perhaps even be thought of as a vindication of Hilbert's approach.

Hilbert's intentions are more clearly seen from his criticisms of his predecessors than from his constructive suggestions. In Hilbert (1922, p. 162) he criticizes Dedekind and Frege because they have operated with general concepts.

Frege has tried to find a foundation for the theory of numbers in pure logic, Dedekind in set theory as a branch of pure logic, neither has reached his goal.

The reason is that they have operated in abstract terms with extensions and intensions of concepts, which according to Hilbert is insufficient and dangerous. In the place of such general concepts as the principal objects of a mathematician's attention Hilbert wants to put

certain discrete extralogical objects which are presented intuitively as immediate experiences before all thought.

These are obviously individuals in contradistinction to general concepts. Hilbert is therefore advocating as a basis of mathematics and logic a study of concrete objects that we can grasp directly and operate with. It does not seem to me far-fetched to call such a vision of the foundations of mathematics combinatorial.

Admittedly the sense of "combinatorial" in which I am using the term here is somewhat vague. I believe nevertheless that there are enough connections between what I have in mind and combinatorics in the generally acknowledged sense, not only to make my usage defensible but to make it interesting. Combinatorial theory, especially Ramsey theory, has turned out to have uses in many different branches of mathematics, perhaps especially striking in logical number theory (see here Graham, Rotschild, and Spencer 1990, Ch. 6). In a different direction, infinitary combinatorics, especially partition theory, is playing an increasingly important role in set theory, as witnessed by titles like *Combinatorial Set Theory* (Erdös et al. 1984; cf. Kleinberg 1977; cf. Williams 1977). It is perhaps symbolic that Ramsey theory started with his 1930 paper which was devoted to the *Entscheidungsproblem* of logic.

The sense of "combinatorial" presupposed here comes close in meaning to "first-order", in that first-order reasoning involves only structures of individuals (particulars), independently of the existence or nonexistence of any properties or relations or sets or classes, or any other higher-type entities. Unfortunately, the term "first-order" is too closely associated with the received attempts to systematize this basic part of logic. As has been shown in this book, these attempts do not exhaust all genuinely first-order reasoning. Indeed, IF first-order logic is neglected in the usual treatments of "first-order" logic.

The same point can be illustrated by means of historical example. I have suggested that one of the leading ideas of David Hilbert's thinking was to base all mathematics on combinatorial realities. It is therefore in keeping with the link between what I here call combinatorial reasoning and first-order logic that one of the crucial steps in the crystallization of first-order logic was taken in Hilbert and Ackermann (1928; cf. Moore 1988). But Hilbert was not satisfied with that systematization. He tried to make more explicit the nature of quantifiers as embodiments of certain choice functions by means of his epsilon-calculus. As noted above, Hilbert hoped in this way to justify the apparently set-theoretical axiom of choice. This attempt can be thought of as another way of trying to implement the same ideas as are systematized in game-theoretical semantics. Moreover, Hilbert was not completely happy, either, with the treatment of negation that is codified in the usual systems of first-order logic. For such reasons, it might be misleading to refer to his logical preferences as first-order ones. I can only hope that the term "combinatorial" does not prompt other misunderstandings.

It must be emphasized that my reduction of mathematics to IF first-order logic is not a translation. One way of describing what is happening is to consider a given higher-order axiom system T. It specifies a class of (higher-order) structures as its models. The reduction I have described means that these structures can be captured by means of extended IF first-order formulas. However, they do not appear as classes of models of given formulas, but as structures embedded in such models.

In brief, for each mathematical axiom system A_1 expressible in terms of standardly interpreted higher-order logic, I have shown how to specify an axiom system A_2 formulated in terms of extended IF first-order logic in which A_1 admits of a relative interpretation. In this sense, the reduction I have sketched is a reduction by relative interpretation. Such reductions are sometimes called *conceptual* reductions (see, e.g., Feferman 1993, p. 148). As was indicated in the preceding chapter, such reductions are not very popular these days. The reason is that such reductions all too often are not what Feferman calls *foundational* reductions. Speaking of relative interpretations, he writes (Feferman 1993, p. 148):

... a familiar example is that [interpretation] of Peano Arithmetic PA ... in Zermelo–Fraenkel set theory ZF..., where the natural numbers are interpreted as finite ordinals. This is a *conceptual reduction* of number theory to set theory, but not a *foundational reduction*, because the latter system is justified only by an uncountable infinitary framework whereas the former is justified simply by a countable infinitary framework.

I am here concerned with justification in a sense that is somewhat different from that of Feferman's. My focus is not the contrast between countable and uncountable infinity, but the problem of the existence of higher-order entities. If this is one's focus, then the reduction I have shown how to carry out is indeed a foundational reduction with a vengeance. The reduct is more fundamental in an important respect than the theory to be reduced. Indeed, in the reduction all questions concerning the existence of higher-order entities are replaced by questions concerning the truth of extended IF first-order sentences (in case we want to know if the axioms are true) and by questions concerning the validity of unextended IF first-order sentences (if we are interested in matters of theoremhood). If this is not a foundational reduction, it is hard for me to think what can be.

These observations can be illustrated and put into perspective by relating them to the so-called Skolem paradox on the one hand and the distinction between the two functions of logic in mathematics (cf. Chapter 1) on the other.

The Skolem paradox is less a paradox than an insight into the limitations of ordinary first-order logic. It is prompted by the Löwenheim-Skolem theorem which says that if a first-order sentence is satisfiable in an infinite model it is satisfied (true) in a *countable* model. What follows from this metatheorem is that first-order languages are not a satisfactory vehicle for discussing noncountable sets, for no first-order formula or axiom system can distinguish a noncountable set from countable ones. In the preceding chapter, it was indicated how these limitations apparently are manifested in (first-order) axiomatic set theory. Such applications have probably contributed to the air of paradox about the Löwenheim–Skolem theorem. This reputation is in reality thoroughly undeserved.

As was pointed out in Chapter 3, the Löwenheim–Skolem theorem holds in IF first-order logic. How, then, can it serve as the universal medium of mathematical problems, as was suggested? What I have shown is that any mathematical theory finitely axiomatizable by means of higher-order logic can be formulated (in the sense explained above) as a contradictory negation of an IF first-order sentence, in such a way that any putative theorem can be taken to be an ordinary first-order sentence (or an unextended IF first-order sentence). Then the claim that it really is a theorem is tantamount to claiming that an unextended IF first-order sentence is valid (logically true).

A mathematical theory formulated in this way by means of extended IF first-order logic can very well have only uncountable models, for the Löwenheim–Skolem theorem does not apply to extended IF logic. At the same time, as was seen, questions of theoremhood can be reformulated so as to pertain to the validity of unextended IF formulas only.

This analysis of the situation can be thought of as a resolution of the Skolem "paradox". It shows that (and in what sense) mathematical theories can deal with uncountable structures even though their logic – at least the logic of theoremhood – can be handled in a logic that admits of the Löwenheim–Skolem theorem. In other words, we can now see how mathematical theories can deal not only with sets but even with uncountable sets, and yet mathematical reasoning is essentially combinatorial.

There is also something of a correlation here with the first two functions of logic in mathematics as outlined in Chapter 1. The descriptive function requires extended IF first-order logic whereas the consequence relations (questions of theoremhood) turn on the validity of unextended IF first-order formulas. These observations will nevertheless be placed in a somewhat different light in the last section of the next chapter.

One important facet of the possibility of using IF first-order logic as the framework of mathematics is that then the metatheory of mathematical theories can be developed by means of the same logical tools as were used to develop these theories in the first place.

In brief, the logic of mathematical theories can be self-applied. The mathematics of metamathematics (cf. Rasiowa and Sikorski 1963) can in principle be taken to be the same mathematics as it serves us to metatheorize about.

This can be illustrated by considering the most important specific results of self-applied logic and mathematics. For instance, my qualified reduction of a large part of mathematics to IF first-order logic can be cast into sharper relief by our other results, especially by our truth definition for IF first-order languages. The possibility of such a truth definition means that important aspects of an IF first-order language can be studied by means of that very language. Even though this possibility does not cover everything in the metatheory of the mathematical theories in question, it is in principle highly interesting because it has often been thought that such self-study is impossible. More specifically, this impossibility is thought of as being implied by Gödel's results. For instance, van der Waerden (1985, p. 157) writes:

> From this Hölder concludes that it is impossible to comprehend the whole of mathematics by means of a logical formalism, because logical considerations concerning the scope and limits of the formalism necessarily transcend the formalism and yet belong to mathematics. This conclusion is fully confirmed by later investigations of Kurt Gödel.

Such claims are at best seriously misleading. Gödel did not prove the descriptive incompleteness of any mathematical theory *an sich*, without reference to some underlying logic or the other. On the contrary, we can see that some of the most important facts about a suitable logical language, namely, the truth-conditions of its sentences, can after all be formulated in the language itself.

But suppose I am asked to respond to Tarski's letter to Neurath of September 7, 1936, quoted in Chapter 1, and to say whether the problem of a universal language had been definitively disposed of. What do my results imply concerning the possibility of a universal language? Is such a language possible? I am tempted to answer, yes and no. However, in this case yes is closer to the truth than no. What we cannot have is a single universal language that need not be extended further. The first truth definition that I explained in Chapter 6 presupposes that any actual finite language system must be capable of being extended by adding new individual constants. Moreover, and even more importantly, there is no hope of formulating a single language in which all mathematical theories could be discussed, a language in which the kind of "model of all models" could be described that set theory was originally thought of as being.

But even though we cannot have a universal language of mathematics, we can have a universal logic. As such a logic, extended IF first-order logic, has been seen to fill the bill. As has been shown, any normal mathematical theory can in principle be formulated in a language whose logic is extended IF first-order logic. If we want to formulate, study and apply another mathematical theory, we can formulate it in a similar but different language. Moreover, we can pool any finite number of such languages together by the simple device of relativization. (It is worth pointing out that the painlessness of such a unification is due to the fact that the theories in question are *first-order* theories.) After the merger we can study the relationship of the different theories in question in the resulting richer language.

In so far as such as open and potentially forever growing sequence of bigger and better languages counts as one language, it qualifies as the universal language of mathematics. Admittedly, it is not one language by the letter of most current definitions, and admittedly different mathematical theories occupy themselves with different parts of it. Since this language is open one cannot discuss it as a completed whole, even though at any stage of its evolution the syntax and semantics of the language fragment so far reached can be discussed in the fragment itself.

In other words, I am not advocating a universal language of mathematics in the strict sense of the word. Different branches of mathematics have different primitive notions. Even if some of them, for instance the notion of cardinality, could be defined in purely logical terms, I still do not see any theoretical or practical payoff in doing so. There does not seem to be any point in trying to combine different mathematical theories somehow so as to become parts of some huge super-theory.

What I have argued is that most (and perhaps all) of mathematics can in principle be done by means of one and the same logic, IF first-order logic. Moreover, that logic is a genuine article and not a disguised version of set theory because it is first-order logic and therefore free of the philosophical problems that have beset set theory or type theory. It has the same – and a better – claim to the title of logic as ordinary first-order logic.

Needless to say, I am advocating this purely logical nature of mathematical theorizing as a rational reconstruction calculated to solve philosophical and other theoretical problems. As to practicalities, presumably the most intuitive way of mathematical theorizing would be to carry it out on a second-order level, or maybe with a simple type theory as its logic; in other words, not unlike what is done in general topology. Only when theoretical problems that have to do with the status of higher-order entities that have to do arise will there be a reason to resort to the kind of reduction to IF first-order logic described in this chapter.

The overall conception of mathematical theorizing that evolves from these observations is not entirely unlike Hilbert's idea of the axiomatic method as used for mathematical purposes. This comparison will have to be a *mutatis mutandis* one, however, for many of the features of Hilbert's views have been dropped or modified here. They include for-reaching changes in the nature of the basic logic, and the abandonment of set theory as a framework for mathematical theorizing, and so forth.

My reduction of all mathematical problems to questions concerning the validity of sentences of an IF first-order language has both mathematical and philosophical significance. One kind of mathematical significance is that it shows that practically all mathematical problems are at bottom combinatorial rather than set-theoretical. This implies that the notion of truth applies in mathematical theories. If you look at set theory, especially in its familiar axiomatic dress, you have a theory whose intended models are not clearly understood, so that the choice of stronger assumptions seems not to be guided by questions of truth and falsity but by some vague "intuitions", or else by considerations of mathematical taste and expediency. In contrast, combinatorial problems are clear-cut. Either there exists a structure of a certain kind or else there does not exist one. Either your jigsaw puzzle or tiling task can be completed or else it cannot be. We have in such cases a razor-sharp characterization of the structures whose existence we are speaking of. This provides no reason whatsoever for dispensing with the notion of truth. The search for stronger deductive premises will be guided by one's combinatorial experience.

This can be illustrated by means of examples. For instance, the stronghold of the defenders of the notion of truth in mathematical contexts has always been the structure of natural numbers. We have such a marvelous familiarity with it that elementary arithmetic is spontaneously thought of as dealing with truth and falsity in this given structure (intended model). Calculatory practice does not define this structure – it merely helps to provide insights into the combinatorial structure of this model.

Now we can see that practically all of mathematics can be taken to deal with similar problems. The structures involved are likely to be less familiar than that of natural numbers, but the character of the problems is in principle similar. In a typical mathematical problem, we are dealing with an axiom system and a putative theorem. The combinatorial insights needed to decide the status of that alleged theorem concern the models of the axiom system. They are parallel to our insights into the structure of natural numbers. They are supplemented *ad hoc* by insights into the combinatorial claims made by the particular putative theorem in question.

In a more philosophical direction, the reduction I have examined means that abstract entities are not indispensable in mathematics in the last analysis (pardon the pun). Frege was wrong; mathematics is not a study of general concepts, but of structures consisting of particulars (individuals).

The results so far reached suggest a view of mathematics which is more pluralistic than the traditional idea of mathematics as being done, not only by means of, but within the scope of logic, set theory or type theory. This traditional picture is inappropriate in several different ways. For one thing, there is no reason to think that all mathematical concepts can be defined within one and the same language, be it logical or set-theoretical. Different mathematical theories study structures of different kinds. There is no reasonable hope to be able somehow to integrate the models of all the different mathematical theories into a single set-theoretical super-universe, as the early set-theorists thought they could do.

Furthermore, the results reached in the preceding chapter and more generally in this book do not only show that axiomatic set theory is an ill-conceived and dispensable basis for mathematics. They also show that one of the main rivals of set theory, to wit, higher-order logics, is likewise dispensable. Both set theory and type theory are theoretically unsatisfactory in that they construe mathematical reasoning and mathematical concept formation as involving essentially higher-order entities. In reality, as has been seen, practically all mathematical problems can be construed as problems in IF first-order logic, and all mathematical axiomatizations can in principle be carried out by means of extended IF first-order logic. This makes type theory as fully unemployed philosophically as set theory is.

Do all mathematical theories and conceptualizations fall within the scope of our results? Can they all be captured, in the sense indicated, in extended IF first-order logic? It is not clear what the answer is, partly because it is not clear what is supposed to count as mathematics. The success of extended IF first-order logic in capturing set-theoretical and second-order conceptualizations suggests a positive answer. There nevertheless seem to be ideas even in perfectly familiar mathematical theories that transcend the purview of the results discussed here. Hence a more appropriate question might be: How far do the results obtained in this chapter extend? The realm of mathematics is a free country. There is no way of anticipating what kinds of conceptualizations mathematicians might decide to use. In the next chapter, I will in fact explore a different and novel perspective on the foundations of logic and mathematics. Hence it is impossible to say anything absolute here. By and large, it nevertheless seems that the observations made in this chapter can be extended further in several important directions. In the next chapter, it turns out that there is an eminently natural constructivistic approach to logic that facilitates a line of thought similar to the one carried out here.

Furthermore, what is probably the most striking prima facie exception to what I have said in this chapter can also be shown to be amenable to treatment on my terms. Even though an explicit treatment would take us too far afield, this case is important enough to be mentioned here.

Probably the best known and most important types of mathematical assumption that do not seem to admit of a formulation in terms of higher-order logic are maximality and minimality assumptions. And probably the best known example of such assumptions is in turn the so-called axiom of completeness which was used by Hilbert (1899) (see Chapter 5). (Strictly speaking, the reference should be to the second through the sixth editions of Hilbert's classic, for the axiom of completeness made its appearance only in the second (1902) edition, and made its exit in favor of an apparently more frugal assumption in the seventh edition in 1930.) This assumption says in effect that the intended models of Hilbert's axiom system are maximal in the sense that mathematical objects cannot be added to them without violating the other assumptions.

Needless to say, maximality assumptions have elsewhere played an important role in mathematics, too, often in combination with minimality assumptions. They open in fact an interesting perspective onto the foundations of mathematics. This perspective is discussed in Hintikka (1993a).

The reason why such maximality assumptions cannot be directly formulated in a higher-order logic is obvious. Saying that a model **M** is maximal apparently involves quantification over individuals and/or sets outside the domain do(**M**) of **M**, while an axiomatic specification of **M** can only be made by means of quantifiers ranging over entities in **M**.

It can nevertheless be shown that such maximality assumptions can also be brought within the scope of the treatment outlined in this chapter, at least in the cases where they typically seem to have been used. If maximality and minimality assumptions are added to ordinary first-order theories, they can easily be handled along the lines indicated in this chapter, thus extending significantly the scope of the treatment proposed here.

With these qualifications, I venture to say that most of the usual mathematics can be done within the framework of IF logic. More specifically, typical mathematical problems can all be reduced to questions concerning the logical truth of different formulas of (unextended) IF first-order logic. In this sense, mathematics can in principle be done on the first-order level, and in *this* sense, mathematics can be thought of as a combinatorial rather that a settheoretical exercise.

My reduction of all mathematical theorizing to IF first-order logic needs to be viewed in a wider perspective. This perspective is provided by the distinction that I made in Chapter 1 between the descriptive and the deductive functions of logic and mathematics. If you review the reduction carried out in this chapter, you will see immediately that it concerns only the descriptive function of logic in mathematics. It concerns the question of what kind of logic is needed to capture and to master intellectually the structures (or classes of structures) mathematicians might be interested in.

In contrast, my reduction does not mean that we can restrict to the first-order level the tools needed for the purpose of dealing with mathematical theories deductively. Indeed, I have apparently strayed unrealistically far from all questions concerning actual logical inferences and hence far from the deductive function of logic. For instance, in ordinary first-order logic, the truth of the conditional $(S_1 \supset S_2)$ is precisely what is needed to move from the truth of S_1 to the truth of S_2. In contrast, the truth of $(S_1 \supset S_2)$ (i.e., of $\sim S_1 \vee S_2$) in IF first-order logic is much more than what is needed to make this move in IF first-order logic.

Hence something more has to be said about the deductive function of logic from the viewpoint of the game-theoretical approach to logic and mathematics. This task will not be attempted in this book, although some light on the matter will be shed by my last two chapters. There, a new aspect of the foundations of logic and mathematics will be examined, namely, the interpretation and implementation of the claims of the constructivists.

Notes

[1] Warren Goldfarb (1989) may be right in defending Russell against Ramsey if we assume Russell's nonstandard interpretation in the first place. But in a deeper sense, it was precisely the nonstandard interpretation that Ramsey was criticizing.

[2] For further information about the scope of Σ_1^1 and Π_1^1 logics, see Moschovakis (1974), especially Chapter 7.

[3] By the theory of type 1 mean theory of *simple* types in contradistinction to Russel's ramified theory of types.

[4] The term "combinatorial" has to be handled with care. What I want to emphasize by the use of the term is the combinatorics of the objects of mathematical theories, and not the combinatorics of the formulas dealing with those objects. This emphasis is therefore radically different from the motivation of the use of the same term by Benacerraf (1973).

10

Constructivism Reconstructed

The approach represented in this book has a strong spiritual kinship with constructivistic ideas. This kinship can be illustrated in a variety of ways. One of the basic ideas of constructivists like Michael Dummett (1978, 1993) is that meaning has to be mediated by teachable, learnable, and practicable human activities. This is precisely the job which semantical games do in game-theoretical semantics. These games can be thought of as being a variety of Wittgensteinian language games. Now these very same Wittgensteinian ideas have been one of the main sources of inspiration to contemporary constructivists. In view of this close relationship of my ideas to those of the constructivists, it is in order to ask what relevance the concepts and results reached here might have to the prospects of a constructivistic theory of the foundations of mathematics.

The answer to this question is not immediately obvious. It might seem that the results reached in the earlier chapters of this book entail a virtual *Aufhebung* of all constructivistic approaches to the foundations of mathematics. This loaded Hegelian term is appropriate because it almost looks as if I had perhaps vindicated the constructivistic approach to logic and mathematics by refuting it. As was mentioned, the basic ideas of my approach are very much in the spirit of constructivistic ways of thinking. Yet I have ended up apparently rejecting many of the characteristic tenets of constructivism. Admittedly, I have not aimed at refuting a constructivistic stance *tout court*, and I do not claim to have done so. However, I have apparently realized some of the main ends of the constructivists without committing myself to the consequences which constructivists themselves embrace. On the one hand, a perceptive reader has undoubtedly noted that I have satisfied some of the most characteristic *desiderata* of constructivists like Dummett. In particular, in my approach to meaning, truth is grounded on certain humanly playable language games, just as Dr. Dummett ordered. On the other hand, I have ended up rejecting virtually all the consequences which constructivists hold dear. I have abided by classical mathematics. Admittedly, I suspended, so to speak, temporarily the law of excluded middle. But it was shown that this rejection of *tertium non datur* for the strong negation is an inevitable feature of every natural basic logic and that it therefore has nothing whatsoever to do with constructivism. Indeed, in the preceding chapter it was in effect shown that you can do all of classical mathematics in a language in which *tertium non datur* fails. And to add logical injury to philosophical insult, I showed how to reintroduce unreconstructed contradictory negation with its undiluted law of excluded middle.

Moreover, I have repeatedly criticized the cornerstone of the usual strategy of the constructivists. This strategy amounts to characterizing a constructivistic interpretation of logic and/or mathematics by presenting a nonclassical set of rules of logical proof. This methodology has been used by Heyting (1956) in presenting a formal system of intuitionistic logic as well as by such later constructivists as Dummett and Prawitz. It was shown earlier that such changes in the rules of logical proof cannot be what the constructivists are really trying to do.

Moreover, the rules of semantical games should likewise be acceptable to a constructivist. In order to verify an existential sentence $(\exists x)S[x]$ I have to find an individual b such that I can verify (win in the game played with) $S[b]$. What could be a more constructivistic requirement than that? Likewise, in the verification game $G((S_1 \vee S_2))$ connected with a disjunction $(S_1 \vee S_2)$, the verifier must choose S_1 or S_2 such that the game connected with it (i.e., $G(S_1)$ or $G(S_2)$) can be won by the verifier. Again, there does not seem to be anything here to alienate a constructivist.

Also, the rules of my semantical games can be taught and learned. Two human beings can play these games against each other. Indeed, this is how Peirce thought of his semantical games (connected with quantifiers) as being played (see Hilpinen 1983). Semantical games are also finite in that each one comes to an end after a finite, pre-

dictable number of moves. There is no opening here for a constructivistic criticism of game-theoretical semantics.

At first sight, the approach represented here might seem to be
committed to nonconstructive assumptions in other ways. For
instance, the main vehicle of my frequent trips between first-order
and second-order logic has been the principle of choice, which is
often considered a paradigm example of a nonconstructive assumption. For I have used as a second-order translation of any first-order
sentence S of IF (or ordinary) first-order language the sentence that
asserts the existence of the Skolem functions for S. (Plus, of course,
the functions which are needed to take care of dependent disjunctions.) Now even in the case of a simple sentence of the form

$$(\forall x)(\exists y)S[x, y] \tag{10.1}$$

its second-order counterpart is

$$(\exists f)(\forall x)S[x, f(x)] \tag{10.2}$$

But the equivalence of sentences like (10.1) and (10.2) is a form of the
principle of the choice and hence seems to involve a heavy dose of set
theory. In Chapter 2 it was nevertheless shown that this apparent use
of the axiom of choice is harmless, being merely an implementation
of the general idea of game-theoretical semantics. What is more,
I will show later in this chapter that such uses of the principle of
choice can be vindicated even from a constructivistic viewpoint.

All this amounts to a serious indictment of the usual ways of
implementing constructivistic ideas. However, this does not imply
that constructivism as such is wrong. Indeed, my game-theoretical
approach helps us to see what the true nature and the true prospects
of constructivism are. This includes changes in the logic and semantics which I have expounded and used in earlier chapters. What is
more, we obtain a new kind of putative argument for a constructivistic approach to logic and the foundations of mathematics. These new
opportunities for a constructivist are based squarely on the basic
ideas of game-theoretical semantics. In the light of these ideas, it can
be seen that a nonstandard interpretation of logic and mathematics
cannot be implemented merely by modifying the definitory rules of
the game of logical proof, or much less by modifying the rules of my
semantical games of verification and falsification. So what else can
a constructivist do? Here the notion of strategy, which is at the heart
of all nontrivial uses of game-theoretical ideas, once again becomes

crucial. Even if we cannot change the move-by-move rules, we can alter the strategy sets which are available to the players. Now in GTS the truth of a sentence S is defined as the existence of a winning strategy for the initial verifier in the corresponding semantical game $G(S)$. This winning strategy can be represented by a function (or finite number of functions). In the simplest – yet fully representative – case they are functions from positive integers to positive integers. But what kinds of functions? Here my game-theoretical viewpoint facilitates a "transcendental deduction" of the way in which constructivistic ideas in the foundations of logic should be implemented. Or, more explicitly speaking, it shows two different ways which constructivists can take, a narrower and a broader one.

The key to this "transcendental deduction" is unsurprisingly the game-theoretical definition of truth as the existence of a winning strategy for the initial verifier. Here the word "existence" has to be taken literally. It might be tempting to formulate the truth-condition by saying that a sentence S is true if and only if the initial verifier "has" a winning strategy in the game $G(S)$. This formulation would be at least seriously misleading. For, in order to reach the classical conception of truth we must stick to the letter of the definition. There may exist a winning strategy for the initial verifier in the abstract sense of the existence of the relevant strategy function without any actual player being cognizant of it, or perhaps even without any player being able to know the winning strategy. This provides an opening to a constructivist. He or she can require that the verifying strategy functions be knowable functions in the sense that an actual human player can play the semantical game in question in accordance with them.

This idea can be interpreted in at least two different ways. Either of them leads to an interesting implementation of constructivistic ideas. In this chapter I will consider first one particular interpretation of the requirement of playability and leave the others for the next chapter. The demand of playability might seem to imply that the set of the initial verifier's strategies must be restricted. For it does not seem to make any sense to think of any actual player as following a nonconstructive (nonrecursive) strategy. How can I possibly follow in practice such a strategy when there is no effective way for me to find out (or perhaps even know) in general what my next move will be? Hence the basic ideas of the entire game-theoretical approach apparently motivate an important change in the semantics of our first-order languages (independence-friendly or not) and in their logic. The resulting semantics is just like my earlier game-theoretical semantics, except that the initial verifier's ("myself's") strategies are restricted to recursive ones. This is a perfectly well-defined change. It leaves the notation used here completely unchanged (independently of whether the slash symbol is present or not). It also leaves all the game rules (rules for making moves in a semantical game) untouched. Hence it represents an unusual and subtle kind of change in our semantics and our logic. Its result will be called a *constructivistic game-theoretical semantics.*

The change involved in the transition to the new version of GTS is motivated by precisely the kind of argument which appeals to constructivists, and which according to them ought to appeal to everybody. For the basis of my argument was the requirement that the semantical games that are the foundations of our semantics and logic must be playable by actual human beings, at least in principle. This playability of our "language games" is one of the most characteristic features of the thought of both Wittgenstein and Dummett.

I will postpone to a later part of this chapter a discussion whether the constructivistic GTS can really be adequately motivated by the line of thought just outlined. Meanwhile, I will take the constructivistic GTS as an implementation of the constructivistic approach to logic and foundations of mathematics and study its properties. Naturally I am especially interested in the question as to which of the results concerning the foundations of mathematics reached in the preceding chapter remain valid. The following are among the most salient points that can be made here:

(i) The motivation for a constructivistic GTS sketched above might at first seem to be far removed from the intuitions of the intuitionists and other constructivists. It is nevertheless easy to see that our motivation is in reality very close to that of the constructivists. This can be seen by examining what the restriction to recursive strategies amounts to in particular cases. For instance, in (10.1) it means that (10.1) is true only if the truth-making value of y can be calculated from any given value of x by some Turing machine. But this is very much like what the constructivistics require in stipulating that all existential statements must be effective ones. Likewise, a dependent disjunction like

$$(\forall x)(S_1[x] \vee S_2[x]) \tag{10.3}$$

will now be equivalent to

$$(\exists f)(\forall x)((S_1[x] \& f(x) = 0) \vee (S_2[x] \& f(x) \neq 0)) \tag{10.4}$$

where f is a recursive function. But this means that the choice of the truth-making disjunct can be made effectively, by calculating the value of $f(x)$ for the given argument x. And what this means is that the choice between disjuncts in (10.3) admits of a constructive decision principle.

If there is an apparent difference between constructivistic GTS and the ideas of *soi-disant* constructivists, it is due to a mistake of theirs concerning the way in which constructivistic ideas should be implemented. They apparently think that such an implementation must take the form of changing the classical definitory rules of the "game" of logical and mathematical proof. What such proofs establish is logical truth or logical consequence, and not its material truth which is the real issue here. Admittedly, the rules of logical proof have to be changed when we move from classical first-order logic to constructivistic logic of the kind explained here. However, the requisite changes are governed by principles quite different from the ones constructivists suppose. Later, it will be seen that the changes intuitionists make in the rules of logical proof are not the right ones anyway. Indeed, in proofs serving to establish logical truth, it scarcely even makes sense to require that an existential statement like $(\exists x)S[x]$ be constructive in the sense of always giving me a recipe for buttonholing a case in point. In such a proof, we are not arguing about the truth, for example, of an existential statement $(\exists x)S[x]$, but about its possible truth. For such a purpose, what can possibly be wrong with my saying that I choose to argue about an arbitrary specimen of the xs satisfying $S[x]$? In constrast, when it comes to the ordinary material truth of existentially quantified formulas, especially dependent ones, the constructivistic requirements suddenly begin to make sense. Hence the constructivists' concerns about the rule of existential instantiation in proofs of logical truth are misplaced.

(ii) The step from the ordinary GTS to the constructivistic one makes a difference already to ordinary first-order logic. It was shown a long time ago that there are arithmetical sentences that are satisfiable in the domain of natural numbers but which do not have any recursive Skolem functions to satisfy them (see Kreisel 1953, Mostowski 1955). This means that when we move to constructivistic GTS there will be fewer arithmetical sentences true in the domain of natural numbers than before.

The same restriction also lends a new interpretation to IF first-order logic, not just to ordinary first-order logic. In both cases, the consequences of the new interpretation have to be studied separately.

(iii) It is natural to define the falsity of a sentence in constructivistic GTS as the existence of a recursive winning strategy for the initial falsifier, that is, a recursive strategy which wins against any strategy (recursive or not) of the initial verifier. If so, there will be sentences of ordinary first-order logic which are neither true nor false in certain models. Thus, constructivists are not entirely wrong in focusing on the failure of *tertium non datur* as a possible symptom of a constructivistic approach. Unfortunately for them, the law of excluded middle fails in perfectly classical nonconstructivistic logic as soon as informational independence is allowed. Hence the failure of the law of excluded middle is not a sufficient condition of a constructivistic logic, in our sense of constructivism. The failure of *tertium non datur* is a fetish, and not a touchstone, of constructivism.

(iv) Now we can see another aspect in which traditional constructivists are barking up the wrong rules. It is not the definitory rules of semantical games that need to be changed, even if we adopt a constructivist stance. It was seen earlier that the rules of logical proof for a constructivistic logic will have to be different from what they are in classical logic. With respect to the definitory (move-by-move) rules of semantical games, the situation is different. When we step from the classical first-order logic to constructivistic logic, no changes whatsoever are needed in the rules that govern the way the two players make their moves. Thus the step from classical to constructivistic logic is as if we could change chess into a different game without modifying its usual rules for making moves, for mating and so forth, merely by putting restrictions on the strategies chessplayers may use. This invariance of the concrete rules for playing a semantical game shows that classicial and constructivistic logics are on a par when it comes to teachability, learnability and the rest of actual playability, apart from questions of strategy. If it is possible for human beings to learn the rules of constructivistic logic, then it is equally possible for them to learn the rules of classical semantical games. It is perfectly possible for actual human beings to learn and master the definitory rules (step-by-step rules) of semantical games which give rise to classical logic. The same holds for my constructivistic logic. It is therefore beside the point for the constructivists to try to alter any move-by-move rules of semantical games. What is at issue here, and what is much more important conceptually, is the question as to what strategies we actual human beings can use in semantical games. If constructivism is to be justified, it is here that its justification must lie.

This point can be generalized. It is in principle possible to think of some automaton or other as following *any* consistent set of definitory rules of a game, or at least structurally similar rules as long as the applicability of those rules is mechanically decidable, as they are in semantical games and in logical proofs. Hence it is very hard to think of any reason why *any* game with decidable definitory rules could not be playable in principle. Yet some constructivists are in effect trying to argue that the classical games of theorem-proving and/or verification, even though they have perfectly well-defined decidable definitory rules, are not playable by human beings. They are accordingly trying to prove the impossible. Instead, they ought to have concentrated on the strategic rules of the relevant games right from the beginning.

(v) In any case, constructivism in my sense is not aiding or abetting antirealism in any size, shape or form. Many of the *desiderata* that antirealists have aimed at were built right into game-theoretical semantics, but they did not in the end have the antirealistic conclusions that philosophers like Dummett have expected them to have. The constructivistic interpretation outlined above does not change the situation very much. For instance, constructivistic truth-conditions are completely on a par with classical (game-theoretical) ones. One can perhaps try to argue that any game-theoretical truth-condition is an antirealist one in that it involves games playable by human beings. But what matters are the rules of semantical games, and not the psyche, the epistemic state or the cognitive capacity of the players. The truth of a first-order sentence in a given model is a combinatorial fact about this model. Whether this fact obtains or not is independent of whether any human being (or any robot, for that matter) ever plays the relevant semantical games.

(vi) The ideas which I have codified in my constructivistic logic and semantics are so natural that it is no big surprise that they are not completely news to mathematicians. What is new is the general formulation of the ideas involved. One particular illustration is worth mentioning here. Semantical games have been used in the form of what are known as Diophantine games in number theory. They were introduced by James P. Jones (1974). A handy exposition of their main features is found in Matiyasevich (1993). They are straightforward instances of the semantical games relied on and discussed in this book, played with number-theoretical formulas on the domain of natural numbers. The reason why they are relevant to the present chapter is that the Diophantine games defined by relatively simple number-theoretical equations exemplify the difference between constructivistic and classical games discussed here, in that in relatively simple games there exist winning strategies for one of the players, but no recursive ones. Indeed, this is a reason why Diophantine games are used in studying questions pertaining to the effective solvability of Diophantine and other equations. Other games where similar nonrecursivity phenomena are encountered are studied by, among other authors, Rabin (1957).

Speaking more generally, most of the entire study of the effective solvability of mathematical problems, such as the study of Hilbert's tenth problem, can be cast in the form of a study of the constructivistic truth of mathematical statements in the sense involved here.

Incidentally, such illustrations also serve to drive home the objectivity of the basic notions of constructivistic GTS. By and large, these studies show that the law of excluded middle will fail for surprisingly simple arithmetical statements in the sense that these statements fail to be true or false on the constructivis-

tic interpretation. While this fact does not by itself prove anything, it certainly suggests that this constructivistic interpretation is out of step with our natural ideas about truth as applied
to number-theoretical statements. Indeed, the solvability of an
equation *means* the constructivistic existence of a solution. The
very question of effective solvability would become empty.

(vii) On the philosophical level, it might nevertheless seem as if this
irrelevance of GTS and of constructivism to the problem of
realism could be mooted. For our game-theoretical truth-
conditions are formulated in terms of strategies of verification.
Strategies are essentially functions of a suitable kind. If an
antirealist is also a nominalist, then he or she cannot accept
such entities as functions and cannot quantify over them.
Hence for such a thinker there apparently is no way of defining
realism in the way we have indicated. This loophole is closed,
however, by the truth definition which was explained earlier for
IF first-order languages. It is on the first-order level, and hence
it does not presuppose any higher-order entities like functions.
Hence all reliance on the existence of higher-order entities can
be avoided, and a fortiori the assumption of nominalism does
not strengthen the hand of an antirealist.

(viii) Admittedly, I have used second-order logic liberally to motivate the approach I have proposed and the definitions and
other analyses it gives rise to. This ascent to second-order logic
cannot be faulted for its apparent use of higher-order entities,
for it was seen that these higher-order entities cancel out from
our most important end-products, such as truth definitions.
But it might still seem dubious for procedural reasons. The
second-order translations of first-order sentences are acceptable, so it may appear, only if we presuppose the axiom of
choice, which is often taken to be an archetypically nonconstructive assumption.

We are in this quandary, it seems, even in dealing with the
very simplest sentences with dependent quantifiers. Consider,
for instance, a sentence of the form

$$(\forall x)(\exists y)S[x, y] \tag{10.1}$$

where $S[x, y]$ is quantifier-free. Its translation (looking away
from dependent disjunctions) is

$$(\exists f)(\forall x)S[x, f(x)] \tag{10.2}$$

But the validity of the translation presupposes that (10.1) and (10.2) are logically equivalent. And this equivalence not only presupposes the axiom of choice; it *is* (a form of) the axiom of choice. Hence my entire procedure, involving as it does constant commuting between first-order and second-order levels of the kind illustrated by (10.1)–(10.2), seems to rely on a bridge which cannot bear any constructivistic traffic.

At this very point, the game-theoretical approach puts the axiom of choice in an interesting light. For what is the relation of (10.1) to (10.2) in my constructivistic GTS? In (10.2), the function variable f must of course be now restricted to recursive values (recursive Skolem functions). Furthermore, (10.1) is now true if and only if there is a recursive winning strategy for its initial verifier. Such a winning strategy is partially codified in a Skolem function f satisfying

$$(\forall x)S[x, f(x)] \tag{10.5}$$

But to say that there is a recursive function of this kind is precisely to assert (10.2), constructivistically interpreted. Hence (10.1) and (10.2), as they stand, are equivalent also according to constructivistic GTS.

In brief, if the principle of choice is formulated as an explicit axiom, it remains valid on my constructivistic interpretation. The idea that the principle fails when constructivistically interpreted is due to an optical illusion or perhaps rather, to mix metaphors, to a vicious double standard. When (10.2) is rejected as not being equivalent to (10.1), the function quantifier $(\exists f)$ is in effect interpreted constructivistically but the first-order quantifier $(\exists y)$ classically. Such an incoherent procedure can only create confusion, however. When all quantifiers are interpreted constructivistically, no matter what their level, the principle of choice changes from a daring hypothesis to a mere definition of a constructivistic interpretation of first-order quantifiers. This is the vindication of the axiom of choice anticipated above. It shows that I did not make any nonconstructivistic commitments when I used equivalences like that between (10.1) and (10.2) in my argumentation.

A review of my vindication of the axiom of choice shows that its nerve is simply the game-theoretical definition of truth as the existence of a winning strategy for the initial verifier. This illustrates what was said in Chapter 3 of how the axiom of choice is little more than one particular facet of the game-theoretical approach to logic in general. Ironically, this approach was originally inspired by constructivistic ideas in the guise of Wittgenstein's notion of language-games.

These observations confute those philosophers and mathematicians who have expressed doubts about explicitly formulated versions of the principle (axiom) of choice. They should encourage those latter-day intuitionists who have quietly sought to reinstate the principle of choice. An example of such recantations of the intuitionists' initial criticism of the axiom of choice is found in Dummett (1977, pp. 52–54). Dummett motivates his acceptance of the axiom of choice by borrowing in effect a page of my book, saying that the axiom of choice "is only dubious under a half-hearted platonistic interpretation of the quantifiers." The thrust of my remarks would also have delighted Hilbert, who once expressed (as we saw) the hope that the principle of choice could be shown to be as obvious as $2 + 2 = 4$. In constructivistic GTS it virtually is just that, quite as much as it is in classical GTS. Thus we are ready to put to rest for good all the doubts concerning the axiom of choice in the foundations of mathematics, except for an epitaph to be presented in Chapter 11.

(ix) Almost by the same token, it can now be seen that my constructivistic logic does not offer any aid and comfort to axiomatic set theory. The hopes that were tentatively raised in Chapter 8 turn out to be false. The suggestion was that by interpreting quantifiers constructivistically we could vindicate the intuitively accurate truth predicate for axiomatic set theory which turned out to be classically unacceptable in Chapter 8.

(x) My constructivistic interpretation of logic might at first sight appear not only new-fangled but somewhat strange. It is therefore in order to point out that it has a respectable, albeit only partial, precedent. It is Gödel's interpretation (1958) of first-order logic and arithmetic, known variously as his functional interpretation or the *Dialectica* interpretation, in honor of the forum of its initial appearance.

Gödel's interpretation begins just like mine. Each first-order sentence S is replaced (interpreted) by a second-order sentence $S*$ which asserts the existence of the Skolem function of S (plus, of course, the similar "Skolem functions" of the disjunctions of S). Then the range of the function quantifier which asserts this existence is restricted to recursive functions. This is in effect the same as we have done.

In Gödel's *Dialectica* interpretation, however, something else is also done. In effect, negations and conditionals are translated into the second-order forms by means of nonstandard rules. These rules are chosen in such a way that the translation process can lead us beyond second-order formulas to even higher-order ones. Basically, the rule for negation interprets it as the contradictory negation, and the rule for conditional is chosen so that $\sim F$ and $(F \supset 1 = 0)$ are equivalent.

These two rules can nevertheless be chosen in a different way which is arguably much more natural. If Gödel's functional interpretation is modified in this way, it comes closer to ours (see Hintikka (1993c).

(xi) Gödel's functional interpretation illustrates how our constructivistic interpretation can be extended to higher-order logics. What we need to do is to translate each statement of order n to a statement of order $n + 1$ all of whose existential quantifiers are initial quantifiers of order $n + 1$ (at most). This translation can be carried out in just the same way as in transforming first-order sentences into equivalent Σ_1^1 second-order sentences. Then all functions and functionals are restricted to recursive ones, just as in Gödel's interpretation. Since the axiom of choice is available, much of traditional mathematics remains in force on this interpretation.

(xii) The constructivistic logic envisaged here has certain major advantages over ordinary first-order logic. One of them is a corollary to Tennenbaum's old result (1959) to the effect that the only recursive model of Peano arithmetic is the standard (intended) one. By a recursive model one means here a model in which the basic relations of elementary arithmetic (sum of, product of) are recursive. But they are recursive if and only if the corresponding Skolem functions can be recursive. And to require the existence of recursive Skolem functions is precisely what characterizes the constructivistic interpretation of logic examined here.

In brief, by means of my constructivistic logic, one can formulate a descriptively complete (categorical) axiom system for elementary arithmetic.

From this result it follows of course that one cannot find a semantically complete axiomatization of the constructivistic logic tentatively proposed here. For if it were possible to find such an axiomatization, elementary arithmetic would admit of a deductively complete axiomatization and hence of a decision procedure. This is in fact an example of the kind of tradeoff between descriptive and semantical completeness that was discussed in Chapter 5. It fulfills, at least by way of an example, the promissory note issued in the same chapter to the effect that there are descriptively complete theories of elementary arithmetic (based of course on a semantically incomplete logic) already *at the first-order level.*

(xiii) One of the advantages of my constructivistic interpretation of IF first-order logic is that it tightens up the correlation between (ordinary and IF) first-order formulas and computer architectures explained in Chapter 4. It was pointed out there that this correlation leaves some first-order formulas high and dry in that their Skolem functions are not computable, in which case no computer architecture can be associated with them. What the constructivistic first-order logic envisaged here accomplishes is to require computability of the Skolem functions of a sentence as a necessary condition of its satisfiability. This removes the unassociated formulas from the correlation and thus a correspondence between *all* satisfiable first-order formulas and certain computer architectures. This goes for both ordinary and IF first-order formulas. The details are explained in somewhat greater detail in Hintikka and Sandu (1995).

More generally, the constructivistic interpretation of logic considered here is very much in the spirit of the times, in the sense that it puts a premium on the computability of the mathematical entities we are dealing with. This is compatible with, if not justified by, the practical importance of computers and computation in contemporary life.

After all these specific comments on the reconstruction of constructivistic logic and mathematics considered here, it is time to return to the general moral of my story. One large-scale question that arises here is: What is changed and what remains unchanged when a constructivistic interpretation is adopted? This question has to be answered case by case.

Perhaps the most intriguing fact about IF first-order logic is that the entire Σ_1^1 fragment of second-order logic can be translated into it. By reviewing the translation process it can be seen that this translatability remains valid also when a constructivistic interpretation (in the sense of constructivistic GTS) is presupposed.

This observation implies that what was said in the preceding chapter about the reducibility of most of classical mathematics to IF first-order logic (and what was said about the sense in which this reducibility holds) remains valid in my constructivistic reinterpretation of logic and mathematics.

For instance, the same line of thought as before shows that for each higher-order statement S_0 we can correlate a Π_1^1 statement. The question whether a certain other higher-order statement S_1 logically follows from S_0 is then equivalent to a question concerning the logical truth of a Σ_1^1 statement. And as was seen in the preceding section, this question is tantamount to the question whether a certain IF first-order statement is logically true. All of this remains valid even if all the higher-order variables are restricted to recursive values.

Thus on the constructivistic interpretation, too, most of ordinary mathematics can be seen to be combinatorial in the sense explained in the preceding chapter.

What is not preserved in the transition to the constructivistic interpretation is the truth predicate discussed in Chapter 6. When one tries to reconstruct this predicate, much of the construction goes through without any problems. For instance, we can give a characterization of what is required of a one-place predicate X (of the Gödel numbers of formulas) to be a truth predicate in the same way as before. But in order for this truth definition to work, the truth predicate must be recursive, for otherwise it would not be a value of the existential quantifier $(\exists X)$. Even though the matter needs further study, there does not seem to be any reason to expect that a constructivistic truth predicate is recursive any more than a classical truth predicate.

If so, interesting suggestions offer themselves here. Maybe the idea of the ineffability of truth has another source besides a belief in the universality of language, namely, a belief in constructivism. This would for instance provide the later Wittgenstein a specific reason to believe in the ineffability of semantics.

These remarks barely scratch the surface of the complex of questions concerning the nature of the newly reconstructed constructivistic mathematics – that is, a mathematics using my constructivistic logic as its basis. Further work is needed in this direction. Meanwhile, we have to face squarely the basic question: Is constructivism correct, at least in its reconstructed form? What makes this question poignant here is that at the beginning of this chapter I gave an argument in favor of constructivism. Moreover, it is obviously a telling argument. It can be summed up in the rhetorical question: Can it possibly make sense to play a game with a nonrecursive strategy? The only reasonable answer that any man (and as Dr. Johnson might have added here, any woman or any child) can give seems to be that it does not make any sense whatsoever. How can I possibly say that I am playing a game with a fixed strategy when I do not have (even in principle) any way of actually calculating what my next move will be?

In spite of the apparent force of these persuasive questions, they do not close the issue. We are here obviously dealing with problems concerning the basic concepts of game theory. And of those basic concepts the most basic one is the notion of strategy. It would not be much of an exaggeration to say that the mathematical theory of games was born the moment John von Neumann (1928) – or was it Borel? – conceived of the current abstract notion of strategy and began to use it to reduce games to their normal form.

But the seeds of dissension lie in this very notion of strategy. On a closer look, it is seen to involve a sweeping and possibly unrealistic abstraction. This abstraction is codified in the notion of strategy employed routinely by game theorists. This concept is used by them to reduce an entire play of a game to a choice of strategy by each of the players. And that certainly looks like an overabstraction if ever there was one. In chess, a grandmaster may perhaps choose his or her strategy (selection of moves) for the first six, eight, ten or twelve moves, as witnessed by the chess clock. In a World Championship match, the first several moves of a game are sometimes made virtually instantaneously, thus revealing an antecedent partial strategy choice. But after that, even grandmasters have to start creating their strategies "across the board", as the saying goes. In doing so, the player in question uses his or her partial knowledge of the opponent's strategy, as it is revealed in the play so far. This can (and should) be generalized to all games of strategy. In all but the very simplest ones, players do not choose their strategies once and for all, but create their strategies in the course of a play of the game in question. This strategy is properly so-called in a very real sense, for it typically involves considerations concerning what my opponent would do if I made a certain move, what I could do in response, and so on. (Of course, the strategy in question is according to the strict game-theoretical definition at best only a partial one.) But if a player's strategy is created in this way, there is no good reason to assume that the resulting eventual strategy is recursive. There even exist probabilistic criteria which tell you, given an initial segment of the values of an integral-valued function, how likely it is to be a nonrecursive one. Applied to the strategy function of a player of a semantical game, such criteria might even justify an empirical finding (albeit only a probabilistic one) to the effect that a player is employing a nonrecursive strategy in a play of a semantical game.

Ironically, the line of thought proposed in this chapter on behalf of constructivists can be turned into an argument against them. It was just seen that there is nothing incoherent about a human being playing a game without an antecedently decided strategy, recursive or not. This is just like a chessplayer creating his or her strategy across the board. What one cannot do is to program an automaton to play a game without a recursive strategy. Even though many constructivists emphasize the role of human thinking and human constructions in the "game" of mathematics, the main argument on their behalf virtually amounts to arguing that since robots have to be constructivists, we humans, too, have to be constructivists. For the only strategies that a digital automaton can be programmed to play in accordance with are the recursive ones. This is precisely what the proconstructivist argument sketched above has alleged to be the inevitable human predicament. The constructivists are in effect imputing to human beings those very limitations that characterize digital computers. It is thus the classical mathematicians who have more faith in human creativity than constructivists, contrary to the occasional claims of the latter.

It is thus difficult to decide whether or not the constructivistic limitation of verificatory strategies to recursive ones is sound or not. What has been seen nevertheless shows that the constructivistic answer cannot be accepted simply on the basis of its greater realism, as far as the actual playing of a semantical game is concerned. On the contrary, the restriction to recursive strategies arguably overlooks the entire realistic possibility of an across-the-board construction of strategies in semantical games. However, further thought is perhaps needed here.

My reconstruction of constructivism brings to the open another subtle problem. The requirement that constructivistic language games must in principle be actually implementable can be turned against the constructivists. For this purpose I only need to look at semantical games from the vantage point of a referee or perhaps a Quinean jungle linguist, rather than that of a player. How can a referee decide on the only basis that he or she has, to wit, on the basis of a player's behavior, whether the player in question is breaking the rules of the game by using a nonrecursive strategy? In a concrete implementation of a semantical game, such a determination must be made on the basis of a finite number of moves. But any finite sequence of moves is compatible with a recursive strategy. Hence a restriction of the use of recursive strategies seems unenforceable.

This problem might at first seem like mere quibbling. In practice, as was noted earlier, it is by no means impossible to draw at least-probabilistic inferences from a player's moves to the strategy that he or she is using. In Las Vegas casinos, pit bosses can spot a "counter", that is, a player who uses a system at blackjack tables. A theoretical basis for such inferences is provided by the studies of Martin–Löf (1966, 1970) and others of the statistical properties of recursive sequences of numbers. Such statistical properties naturally manifest themselves in finite sequences of events (see here Fine 1973, Ch. 5 and Schnorr 1971).

But the comparable implementation of the recursivity restriction in semantical games can only be probabilistic, and not strict. This introduces a new and unfamiliar complication in that the outcome of a play of the game will depend on how the rules of the game are enforced. Such games may or may not be a territory on which game theorists fear to tread, but so far they do not seem to have done so in a way that would help us here with our present problems. Hence the conceptual problems involved in the implementation of a recursivity restriction do cast a shadow on constructivistic philosophies of logic and mathematics.

There is another kind of reason against restricting Skolem functions to recursive ones. Such a restriction would make it extremely hard to develop a mathematics that could serve the purposes of physical theory. It has been shown that certain perfectly deterministic systems in classical mechanics exhibit nonrecursive behavior (see Ekeland 1988, pp. 59–61). Such a behavior would be impossible or at the very least extremely awkward to handle in mathematics where functions are restricted to computable ones.

It is in any case important to realize what precisely is involved in the transition from a classical first-order logic (IF or not) to the corresponding constructivistic logic. In a sense, both logics are arguably constructivistic. Both kinds of logic are grounded on certain "language games". Moreover, these language games are the same on the classical and on the constructivistic side of the fence, in the sense that precisely the same moves are admissible in the two cases and that winning and losing are defined in the same way. Hence, *pace* many constructivists, if constructivistic logic is teachable, learnable, recognizable from peoples' behavior, and actually practicable by human beings, then so is the corresponding classical logic. What makes the difference is the use we make of the language games in question for the purpose of defining a notion like truth.

Even though the constructivistic logic developed here is an eminently natural and interesting way of implementing constructivistic ideas, it does not match everything that intuitionists and other constructivists say about their logic. However, it is not obvious what such discrepancies prove. In some cases at least, they serve to illustrate the confusions of the constructivists. For instance, it is clearly in accordance with the intentions of the constructivists that more is required of the constructivistic truth of a mathematical theorem than of its classical truth. This idea is implemented by the constructivists discussed here. Yet the notion of constructive truth so implemented is not in agreement with the fine print of all of the so-called constructivistic theories of logic and mathematics.

For instance, when the constructivists' aims are realized along the lines proposed here, we cannot eliminate the undecidability of mathematical problems in the way that some constructivists wanted to do. However, to my mind this result only goes to show that the constructivists' aims were incompletely analyzed. Decidability is closely related to deductive completeness, and such completeness is shown by Gödel's results (and by the role of IF first-order logic as the basic logic of mathematics) to be a pie in the sky. What remains of the constructivists' motivation is therefore an emphasis on the epistemic element in mathematical theories. But this idea can, and should, be implemented in an entirely different way. For one thing, the epistemic element should be dealt with above the board and not merely tacitly. In the next chapter, I will show how this can be done in a simple and elegant way.

In sum, the claims of the constructivists to be able to give us a satisfactory general perspective on logic and mathematics seem to me seriously flawed. However, if we lower our sights somewhat, the situation can be seen to change. If we restrict our attention to the deductive function of logic in mathematics, then constructivistic ideas will suddenly be seen to fill a genuine need.

It was noted at the end of Chapter 9 that the game-theoretical approach apparently has little to contribute to, or even to say about, the deductive function of logic. This admission might have struck the reader as being premature. Indeed, the game-theoretical conception of truth might prima facie seem to show precisely what there is to be done for the purpose of facilitating the deductive task of logic. What is needed is an analysis of what it takes to infer the truth of a sentence, say of

$$(\forall z)(\exists u)S_2[z, u] \tag{10.6}$$

from the truth of another one, say of

$$(\forall x)(\exists y)S_1[x, y] \tag{10.7}$$

For simplicity, I will in this example disregard the role of all quantifiers and connectives in S_1 and S_2.

The truth of (10.7) means that there is a winning strategy for the initial verifier in the correlated game. If the simplifying assumptions just indicated are made, this means the existence of a function f such that

$$(\forall x)S_1[x, f(x)] \tag{10.8}$$

Similarly, the truth of (10.6) means that there exists a function g such that

$$(\forall z)S_2[z, g(z)] \tag{10.9}$$

We can therefore get from the truth of (10.6) to the truth of (10.7) by means of a function which takes us from a winning strategy in the game correlated with (10.7), that is, from the function f, to a winning strategy in the game correlated with (10.6), that is, to the function g.

This means obviously the existence of a functional Φ that yields g as a function of f. This can be expressed by the third-order sentence

$$(\exists\Phi)(\forall f)[(\forall x)S_1[x, f(x)] \supset (\forall z)S_2[\Phi(f), z] \qquad (10.10)$$

More generally, we can in our metalogical notation associate with each formula two metalogical variables – one for the strategy function of each of the two players. If the strategy functions of the initial verifier and the initial falsifier associated with the premise (say A) are ξ and η and those associated with the conclusion (say B) are ϕ and ψ, we can then say in general that the inference-ticket conditional that mediates the passage from the premise to the conclusion is

$$(\exists\phi)(\forall\xi)[(\forall\eta)A(\xi,\eta) \supset (\forall\psi)B(\phi(\xi), \psi)] \qquad (10.11)$$

In ordinary higher-order logic this would be equivalent to

$$(\exists\phi)(\forall\xi)(\forall\psi)(\exists\eta)[A(\xi,\eta) \supset B(\phi(\xi),\psi)] \qquad (10.12)$$

and hence to

$$(\exists\phi)(\exists\eta)(\forall\xi)(\forall\psi)[A(\xi,\eta(\xi,\psi)) \supset B(\phi(\xi),\psi)] \qquad (10.13)$$

This, then, might seem to be the kind of conditional that could serve as an inference-ticket for the purposes of logical deduction. Indeed, (10.13) is formally identical with Gödel's interpretation of conditionals.

Unfortunately, in ordinary higher-order logic, (10.13) can be seen to reduce back to ordinary truth-functional conditional, and hence not to give us anything new.

The reasons for this failure of the analysis (10.13) to yield anything useful for the deductive function of logic go back to the very definition of truth as the existence of a winning strategy for the initial verifier in a semantical game. As long as existence is taken in a standard sense, it does not follow that the verifier "has" a winning strategy in the sense of knowing what such a strategy is – or even being in a position to know what it is. In deduction, we are moving from the known truth of the premise or premises to the known truth of the conclusion. In order to handle such matters, it is mandatory at least to restrict the initial verifier's winning strategies to those that he or she can know. And this apparently entails that these strategies can be represented by recursive (computable) functions.

This means using the constructivistic logic outlined earlier in this chapter. In general, while it was not possible to find convincing reasons for adopting a constructivistic standpoint in general in the foundations of logic and mathematics, there seems to be a great deal to be said for adopting it for the purposes of the deductive function of logic in mathematics.

In the particular matter just discussed, we can use our constructivistic logic to get a nontrivial analysis of the conditionals which can operate as a bridge from the known truth of a sentence (in the sense that a winning strategy in the corresponding game is known) to the known truth of another sentence. In the case of a specific sentence like (10.13) expressing the conditional which can bridge such inferences, this constructivistic logic amounts to restricting the functions and functionals in it to recursive ones.

I will not pursue the technical possibilities that are opened up here. However, it is of great interest to examine the general theoretical perspectives that are revealed by my observations.

First, in this way constructivistic ideas can be seen to play a major legitimate role in a mathematician's work in any case. They are needed to understand and to master the deductive task of logic in mathematics. At the same time, this role is different from what constructivistic philosophers like to think. For one thing, the role of constructivistic notions has nothing to do within the meaning of mathematical statements. Meaning is a matter of the descriptive function of logic in mathematics. It is a matter of sentence–model relationships, ultimately a matter of truth definitions. It is not a matter of deductive relationships between propositions.

Second, the analysis of conditionals (10.10), combined with the constructivistic logic outlined above, can be thought of as yielding Gödel's so-called functional interpretation of first-order logic. The line of thought just indicated provides a new theoretical justification for Gödel's interpretation and at the same time shows its roots in the game-theoretical approach to logic.

It may be noted that the Gödelian conditional (10.10) is equivalent to the metatheoretical sentence

$$(\forall \xi)(\exists \phi)(\forall \psi)(\exists \eta)[A(\xi, \eta) \supset B(\phi, \psi)] \tag{10.14}$$

which prima facie involves no step to a higher type in the type hierarchy. But if (10.10) is itself to occur in a conditional, it has to be brought to a form where existential quantifiers precede universal ones. This form is (10.13), which shows that this construal of iterated conditionals leads to higher and higher types. Such iterated conditionals are arguably indispensable in any reasonable treatment of the deductive function of logic, and the ascent in type hierarchy appears to be unavoidable.

This result confirms the suspicions voiced at the end of Chapter 9. The reduction of mathematical theories to the first-order level which was outlined in Chapter 9 applies only to the descriptive function of logic in mathematics. In contrast, an attempt to serve increasingly better the deductive task of logic will lead us in the opposite direction, that is, to higher and higher types, just as in Gödel's functional interpretation.

We can thus see that the distinction made in Chapter 1 between the two main functions of logic in mathematics does in fact matter. The descriptive and the deductive functions are most naturally served by different conceptualizations. It might indeed be salutary for philosophers of mathematics to keep this distinction more firmly in mind than what is customary these days.

The schizophrenic relation of logic to the type hierarchy which has been found is highly interesting philosophically and historically. It is perhaps not too far-fetched to see in Gödel's proof of the deductive incompleteness of elementary arithmetic an indication of the need of having to climb higher and higher in the type hierarchy *for deductive purposes*. This point can be generalized in an interesting way. Mathematics is often thought of as a science of abstract objects and abstract structures. This abstractness is presumably greater the higher we climb in the type (order) hierarchy. Gödel for one entertained a concept of abstraction to which this relationship to type hierarchy can be attributed. In contrast, logic tends to be thought of, at least ideally, as a manipulation of concrete symbols. Such an ideal of logic was represented by the Vienna positivists' idea of the logical syntax, and by Hilbert.

What we have found turns this relationship between mathematics and logic upside down. For the primary descriptive function of logic in mathematics, all that we need conceptually are combinatorial structures consisting of particulars belonging to the rock bottom of the type (order) hierarchy. In contrast, the deductive treatment of such mathematical structures forces us to consider increasingly more and more abstract (higher-type) entities. It may thus be said that the ontology of mathematics can be restricted to the world of combinatorics and of particular nominalistic entities, whereas the deductive technology of mathematics forces us to higher and higher levels of abstraction.

Even though these results have to be handled with considerable caution, they suffice to show how thoroughly misleading, not to say mistaken, our received ideas about logic, mathematics and their interrelations are.

Some of these comments can be given a more personal address. For instance, Gödel not only maintained a Platonistic ontology of mathematics but thought that an ascent to higher and higher abstraction was a way of solving foundational problems. If I am right, the need of abstractive ascent is due only to the needs of actual deductive inferences; Platonistic theorizing, which Gödel advocated in some of his pronouncements, does not require such an ascent.

Also, it is ironic in a historical perspective to see that Hilbert's first and foremost interest was in the descriptive function of logic. This is in evidence not only in his work in the foundations of geometry but perhaps even more clearly in his general remarks on the axiomatic method. The function of a mathematical theory is to capture a structure or class of structures by completely logical means. The development of metamathematics and proof theory which was prompted by Hilbert's foundational project was a mere by-product of his overall enterprise. In a sense that appears from what has been said, proof theory is an enterprise in a direction quite different from Hilbert's central interests.

11

The Epistemology of Mathematical Objects

At this point, it behooves me to reflect on what I am doing and to generalize the questions that I have raised. It was indicated above that the constructivistic interpretation of logic and mathematics outlined in the preceding chapter is not without precedents. The philosophical motivation of the constructivistic interpretation discussed there is new, or at least much more fully articulated than the expressed earlier motivations of similar views. But the technical implementation of my interpretation does not stray very far from Gödel's *Dialectica* interpretation (1958) of first-order elementary arithmetic or from Kleene's realizability interpretation (see Kleene 1952, sec. 82).

However, it is possible to generalize the entire constructivistic interpretation considered here in a way which is more radically new and which opens the door to a deeper motivation of constructivism. This motivation can be seen by criticizing the way I have tried experimentally to present a raison d'être for the one particular version of constructivism in the preceding chapter. This rationale turned on the alleged impossibility of playing a semantical game in accordance with a nonrecursive strategy. It was found unconvincing. A working mathematician might very well resent the tacit limitations that my tentative argument imposes on a mathematician's ability to play a semantical game. Surely a competent mathematician is not restricted in his or her mastery of strategies to such simple ones as to be codifiable by recursive functions – that is, to functions that even a stupid machine can handle. Even if a mathematician cannot program a computer to calculate a function for all argument values, he or she can do all sorts of other things with it. The value of this strategy function can perhaps be calculated for most (or perhaps almost all) of the argument values. The qualitative behavior of the strategy function can be mastered even if it is not computable by a digital automaton. It can even in principle be calculable by an analogical computing device in contradistinction to a digital one. In brief, a nonrecursive function can be mastered intellectually in a way that would authorize us to say that a semantical game is played in accordance with a strategy codified by this function.

This line of thought is clearly related to what was said in Chapter 10 about the possibility of creating a strategy "across the board". It can be used to evade the criticisms that were rehearsed there against such "classical" GTS as dispenses with the recursivity requirement.

But although this line of argument tells against a restriction of strategy functions to recursive ones, it is prima facie compatible with *some* restrictions on those functions. In fact, it is easily turned into a criticism of the classical interpretation of logic and mathematics along the same lines as were followed above, but imposing a weaker restriction on the strategy (Skolem) functions relied on in a game-theoretical approach to the foundations of mathematics. The argument would then be that the verifier can meaningfully be said to play a semantical game in accordance with a strategy only if this strategy is codifiable in a function which can in principle be mastered mathematically. Surely the very idea of strategy as being used by a human being implies a conceptual control of the successive moves made in accordance with that strategy.

But this argument, even though eminently plausible, now seems to be so vague as to be useless. It is not that one cannot think of what might be involved in mastering a function mathematically. The trouble is that one can think of a large number of different ways of understanding this idea of mastery. What I will do here is to focus on what seems to be the greatest common denominator of these different ways and to interpret one's mastery of a function to mean that this function is a *known* one – more accurately speaking, to mean that *one knows which function it is.* At first sight, this might only seem to introduce a *façon de parler;* for where is the real link between this location and our actual notion of knowledge? Even though philosophers have analyzed the notion of knowledge intensively (or at least extensively), it still does not seem to be sharp enough to be useful here.

A look at the historical situation may provide some clues here. So far, I have not sought to make any distinction between constructivism and intuitionism. This might seem *comme il faut* in view of the current usage in which all and sundry constructivistic tendencies are called "intuitionistic". This usage nevertheless fails to do justice to the specific views of the real intuitionists who are following or, at least, are inspired by Brouwer. His views are not necessarily done justice to by calling them constructivistic. Typical constructivistics pay attention in the first place to logical and mathematical behavior – that is, to what can be done in logic and mathematics. The question of the possibility of actually playing semantical games with arbitrary strategies and the question of how different strategies are betrayed by a player's behavior are typical constructivistic questions. In contrast, Brouwer is inquiring into a mathematician's thought rather than into his or her behavior, and he is raising questions about what we know and can know rather than questions as to what we can do. This is connected with the fact that the true intuitionists have not been happy with the development of the notion of recursivity (Turing machine computability or equivalent) as an explanation of the kind of knowability they have in mind. Their reasoning is revealing. It is not enough that a set of equations in fact enables us to carry out the calculation needed to determine the value of a function for each argument value. We must be able to know that it does so.

More generally, no reader of Brouwer's writings (1975) can fail to be impressed by the role of epistemic ideas in his argumentation. An especially telling case in point is Brouwer's characteristic and crucially important technique of "weak counterexamples" (see Brouwer 1908c in Brouwer 1975; Troelstra and van Dalen 1988, pp. 8–16; van Stigt 1990, pp. 252–255). Such a counterexample to a logical or mathematical principle does not show that the principle is false in any usual sense. It shows the unacceptability of the principle by showing that its acceptance implies that "we ought to have certain knowledge... which in fact we do not possess" (Troelstra and van Dalen 1988, p. 11). This, then, seems to be the dividing line between constructivists and true intuitionists: intuitionists emphasize the role of knowledge in mathematics while constructivists put a premium on the effectiveness of constructions and other operations. Constructivists are do-it-yourself mathematicians, intuitionists are know-it-yourself ones. In view of this contrast, it is not surprising that it has been possible to interpret intuitionistic logic in epistemic logic. This result has to be handled with great care, however, for it will be shown later in this chapter that neither one of these two logics is adequate in its present (axiomatic) shape.

In my considered view, intuitionists have done foundational discussion a monumental disservice in not bringing out into the open the epistemic element, which according to their own views there is in mathematical theorems. I also believe that they have done their own ideas a disservice by not so doing. They apparently thought that there inevitably is such an epistemic element, in effect an implied claim to knowledge, in all mathematical statements. If so, their tacit epistemic claims would be commensurable with the propositions of classical mathematicians only if they, too, have been making tacit epistemic claims. However, I fail to find any strong arguments for such commensurability in the writings of the genuine intuitionists. It is perhaps possible to try to view some of Brouwer's pronouncements as steps in this direction. If so, they are scarcely explicit enough to be discussed and evaluated here.

It is not too far-fetched to see in the intuitionists' failure to acknowledge the epistemic dimension of their thinking a confusion between the different language games distinguished from each other in Chapter 2. What gives mathematical statements their meaning are the semantical games on which the truth definition used here is based. However, it is obvious that intuitionists like Brouwer have thought of mathematical activity as a process of actually coming to know mathematical truths and mathematical objects. In Chapter 2 it was seen that these two kinds of activities ("language games") have to be sharply distinguished from each other in the interest of conceptual clarity.

This criticism puts an onus on me to spell out precisely what this epistemic dimension of mathematics alleged by the intuitionists is, and how it can be implemented. At first sight, this might seem a pointless undertaking. As an example, we might ask how the epistemic element might be expressed by means of received epistemic logic. There you find only one epistemic ingredient. It is the "knows that" or "it is known that" operator K. If the knower is to be indicated, a subscript can serve the purpose, so that $K_a S$ is to be read approximately

a knows that S

But such an epistemic ingredient does not seem to help us at all. For in the usual treatments of epistemic logic K_aS is logically true (valid) if and only if the nonepistemic sentence S is logically true. Hence the introduction of the epistemic element does not seem to serve any purpose.

This mildly paradoxical-looking result is not merely a matter of the deductive or inferential relationships between different statements. It is firmly rooted in the model theory of epistemic logic. The one safe conceptual anchor one has there is the idea that knowing something, say, knowing that S, means being able to rule out legitimately all the states of affairs, courses of events and other scenarios in which S fails to be true. No matter what questions can be raised about the legitimacy of such exclusion of scenarios (the philosophers' ill-named "possible worlds"), the resulting epistemic logic remains unaffected.

At this point the aficionados of epistemic logic will undoubtedly expect me to evoke the notorious paradox of logical omniscience. This so-called paradox is a simple corollary to the model-theoretical-conception of knowledge just outlined. As a moment's reflection shows, this conception implies, so it seems, that whoever knows something knows all the logical consequences of what he or she knows.

This diagnosis of the so-called paradox immediately reveals a cure. This cure lies in the study of the processes through which one in fact comes to know the logical consequences of one's premises. This prescription is not hard to use. It can even be handled model-theoretically, as has been shown by the work represented in Rantala (1975) and Hintikka (1975).

This is nevertheless not an occasion to follow up such a line of thought. The main reason is that it is not what intuitionists have had in mind. In the paradox of logical omniscience, in the simplest cases at least, we are dealing with the fact that the application of logical rules of inference does not happen in one fell swoop, but is a matter of step-by-step applications. However, what the intuitionists are doing is not a study of such stepwise procedures. Rather, they propose to change the very rules of logical inference. This is an approach that is entirely different from the ideas of Rantala and Hintikka who focus on the fine structure of inference rule applications. It has to be conceptualized and otherwise dealt with in some other manner.

But how? If the intuitionists were not dealing with *knowing that*, then what was the epistemic element they were implicitly importing into mathematics?

Here the developments which are made possible by the notion of informational independence and which were outlined at the end of Chapter 4 play a crucial role. What was seen there is how to implement an idea which the received epistemic logic does not capture. This is the idea of *knowledge of entities* (objects, things) as distinguished from *knowledge of facts* (propositions, truth of sentences). The latter kind of knowledge is relative to a space of scenarios ("possible world") on which an alternativeness ("accessibility") relation is defined. The former kind of knowledge is relative to a set of "world lines" of cross-identification which define which denizens of the different possible worlds count as identical (manifestations of the same entity). As was emphasized in Chapter 4, the way these "world lines" are drawn is largely independent of the truth-conditions of *knowing that* statements.

This enables me to put forward an interpretation – or rather a rational reconstruction – of the main thrust of the intuitionists' thinking. They were indeed tacitly introducing an epistemic element into mathematics. But this epistemic element is not a knowledge of truths, but a knowledge of entities. It should not be left implicit, either, as intuitionists were doing. It can – and should – be implemented along the lines of the epistemic logic sketched in Chapter 4. This new logic is made possible by the main new conceptual tool introduced in this book, namely, the notion of informational independence.

Before developing this approach systematically, it is in order to recall how it hangs together with the line of thought initiated in the preceding chapter. There the idea was entertained of restricting strategy functions to recursive functions. It was argued that this is far too restrictive a limitation. Instead, it was suggested that the strategy functions should be restricted to *known* functions. This is the restriction which has meanwhile been found to be in agreement with the intuitions of intuitionists. It is the idea that is being implemented here.

Using the notation and the observations explained in Chapter 4, we can express what it means for a function f to be known. The following are the equivalent formulations of this idea:

$$K(\forall x)(\exists y/K)(f(x) = y) \tag{11.1}$$

$$K(\exists g/K)(\forall x)(f(x) = g(x)) \tag{11.2}$$

$$(\exists g)K(\forall x)(f(x) = g(x)) \tag{11.3}$$

$$(\exists g)K(f = g) \tag{11.5}$$

Here K is interpreted in a game-theoretical context as a universal quantifier ranging over epistemic alternatives to the world which the players are considering at the time: the falsifier chooses such an alternative, by reference to which the game is continued.

This suggestion does not yield a unique interpretation of what it means to master a function intellectually, that is, to "know which function it is". On the contrary, the criteria for knowing a function are left almost completely open. Indeed, it is one of the fundamental insights prompted by the semantics of epistemic logic that the truth-criteria of *knowing* + wh-constructions are largely independent of those of *knowing that* statements. In the independence-friendly notation, this means that the truth-conditions of K-statements containing slashed existential quantifiers $(\exists f/K)$, $(\exists x/K)$ are left underdetermined by the truth-conditions of unslashed K- statements. The underlying semantical reason is that they depend on criteria of identification across models (worlds). In contrast, slashed disjunctions $(\vee/K)$ depend on identification conditions of functions only indirectly by being under the scope of universal quantifiers.

This inevitable underdetermination opens the door to several interesting lines of thought. It shows that the basic ideas of intuitionists are open to different specific interpretations, depending on which criteria are applied to our knowledge of mathematical objects. It looks hopeless to try to find a consensus as to which interpretation is the right one. However, this does not spoil the usefulness of the epistemic interpretation schema for the purpose of conceptual clarification. It has already been seen that one can use this epistemic interpretation schema to discuss a number of different intuitionistic and constructivistic ideas. What is more, one can even establish general logical results which are independent of what counts as a function being known.

In order to reach such results, I have to be a little more explicit than I have been so far. I will generalize the intuitionists' way of looking at mathematical theories and theorems into something that might be called an epistemic (or intuitionistic) interpretation of mathematics and logic. Starting from the simplest case, what is the epistemic interpretation of a first-order mathematical statement S? For simplicity, I will assume that S is in a negation normal form. Then we can reinterpret S as "really" meaning

$$K S^* \tag{11.6}$$

where S^* is formed from S by replacing every existential quantifier $(\exists x)$ by $(\exists x/K)$ (and likewise for higher-order quantifiers if they are used) and every disjunction $\vee$ by $(\vee/K)$. This amounts to requiring that all the functions which operationalize existential statements (Skolem functions) are *known* functions.

This new interpretation can naturally be extended to IF first-order logic and to mathematical theories and statements that can be expressed by its means. I will call it the *epistemic interpretation* of logic and of mathematical theories. It is important to realize what is new in it. In the earlier literature on the subject, when epistemic considerations were brought to bear on the foundations, the question explicitly or implicitly raised was: What does it mean to know a mathematical *truth*? Moreover, this question was discussed as if it merely involved *knowing that a* certain *mathematical proposition* holds. What has come up here is an altogether different dimension of the epistemology of mathematics. We are now dealing with the question of our *knowledge of mathematical objects* such as functions. One remarkable but largely overlooked fact here is that these two epistemic dimensions are largely independent of each other. Yet, there is plenty of evidence that in the earlier literature the two kinds of epistemic issues were confused with each other. A case in point is offered by the usual criticisms of the axiom of choice discussed below.

The epistemic interpretation of mathematics offered here is calculated to incorporate both questions of our knowledge of mathematical truths and questions of our knowledge of mathematical objects. However, the precise criteria of the latter kind of knowledge are left open by the epistemic interpretation, which therefore is an interpretational schema rather than a unique interpretation.

Thus a rough-and-ready distinction can be made between constructivistic and intuitionistic approaches to the foundations of mathematics. The former emphasize the role of what a mathematician can in fact do, while the latter emphasize what a mathematician can know. The relationship between the two is a delicate matter, so much so that it is not even clear as to which is the more general view. On the one hand, the constructivistic game-theoretical semantics for mathematics characterized in the preceding chapter can be thought of as a special kind of the epistemic interpretation. It is obtained by stipulating that all and only *recursive* functions are "really known". A part of the interest of the constructivistic interpretation is thus due to the fact that it exemplifies the epistemic interpretation of mathematics.

In this perspective, the epistemic interpretation of logic and mathematics is much broader than the constructivistic one defined in the preceding chapter. For instance, it is quite clear that the hardcore intuitionists rejected the idea that recursivity could serve as the explication of constructivity. Their views therefore fall within the purview of the epistemic interpretation, but not of the constructivistic one. More generally speaking, one important difference between classical and intuitionistic mathematicians is that the former are satisfied with knowing mathematical *truths* while the latter also want to know mathematical *objects*.

On the other hand, the emphasis of Brouwerian intuitionists on the epistemic element in mathematics is seen to lead inevitably to an emphasis on our knowledge of mathematical objects. This emphasis is quite different from an emphasis on our actual ways of coming to know mathematical truths, and in a sense is much more specific.

The epistemic interpretation of mathematics offers a new and powerful tool for future studies of the epistemology of mathematics. A few examples may illustrate its relevance. By means of epistemic logic, the epistemic interpretation can be illuminated by the same means as ordinary logic, namely, through second-order translations. For instance, on the epistemic interpretation

$$(\forall x)(\exists y)S[x, y] \tag{11.7}$$

becomes (if we look away from dependent disjunctions) a shorthand for

$$K(\forall x)(\exists y/K)S[x, y] \tag{11.8}$$

which is equivalent to

$$(\exists f)K(\forall x)S[x, f(x)] \tag{11.9}$$

and to

$$K(\exists f/K)(\forall x)S[x, f(x)] \tag{11.10}$$

But (11.10) is the "translation" (interpretation) of

$$(\exists f)(\forall x)S[x, f(x)] \tag{11.11}$$

Simple though these observations are, they have interesting consequences. What the equivalence of (11.8) and (11.10) shows is that the epistemic interpretation validates the axiom of choice, no matter how a function being known is understood, as long as the epistemic interpretation is applied consistently to all quantifiers. What the critics of the axiom of choice have in effect done, looked at from the vantage point of the epistemic interpretation, is to confuse with each other

$$K(\forall x)(\exists y)S[x, y] \tag{11.12}$$

and

$$K(\forall x)(\exists y/K)S[x, y] \tag{11.8}$$

which is equivalent to

$$K(\exists f/K)(\forall x)S[x, f(x)] \tag{11.10}$$

In other words, the critics of the axiom of choice have tacitly taken the problem of whether one knows the mathematical *statement* $(\forall x)(\exists y)S[x, y]$ to involve also the question as to whether the choice function f (i.e., a certain mathematical *object*) is known or not. This exemplifies the general confusion mentioned earlier.

These observations throw a great deal of light on the philosophical stance of different mathematicians and philosophers. We can for instance see that it is not constructivism but intuitionism, with its emphasis on our knowledge of mathematical objects, that can lead to a rejection of the axiom of choice. Furthermore, from my epistemic vantage point it is highly instructive to see how a prominent defender of the axiom of choice rests his case on the idea that the existence of a mathematical object "is *a fact like any other*" (Hadamard, quoted in Moore 1982, p. 317; emphasis added). The whole claim of the intuitionists is that we are, in assumptions like the axiom of choice, not dealing with knowledge of facts but with knowledge of mathematical objects.

For another instance, Dummett's philosophical position can, in the light of what has been found, seem to be constructivistic rather than intuitionistic according to the distinction adumbrated in this chapter. This is shown not only by Dummett's acceptance of the axiom of choice (see the preceding chapter) but also by his insistence that there is nothing special about the semantics of dependent quantifiers like $(\exists y)$ in

$$(\forall x)(\exists y)S[x, y] \qquad (11.7)$$

beyond the general requirement of constructivistic justifiability or in Dummett's term, assertability (Dummett 1977, p. 451; cf. Troelstra 1977, pp. 154–156).

Thus the epistemic interpretation of mathematics enables us to tell precisely what is valid and what is invalid in the axiom of choice. Likewise, we can see by means of the epistemic interpretation what is acceptable and what is not in another bugbear of the intuitionists, namely, the method of indirect proofs. At first sight, the critics of indirect proofs have once again got hold of the wrong end of a sticky concept. For suppose that a logician has taken the negation $\sim S$ of a statement and showed that it leads to a contradiction, in other words, shown the following:

$$\sim S \supset (S \,\&\, \sim S) \qquad (11.13)$$

But (11.13) is equivalent to S, so that the indirect prover needs only to add "q.e.d." And it does not matter whether or not we prefix K to (11.13).

However, if in KS there are slashed expressions $(\exists x/\mathrm{K})$ or $(\vee/\mathrm{K})$, the negation of S that figures in an indirect proof cannot simply be taken to be $\sim S$ or $\neg S$, for neither of these makes any sense standing alone. The only negation that can be relevant here is $\sim(S/)$ where $(S/)$ is obtained from S by omitting all independence indications (i.e., all slashes with their right-hand accompaniments). Then an indirect proof involves an inference from $\sim \sim (S/)$ to S. Such inferences are not always valid.

For instance, take the epistemic statement

$$\mathrm{K}(\forall x)(\exists x/\mathrm{K})S[x, y] \qquad (11.8)$$

The only negation that can be drafted into the service of an indirect proof is here

$$(\exists x)(\forall y) \sim S[x, y] \qquad (11.14)$$

If this is reduced to impossibility, then what has been reached is

$$\mathrm{K} \sim (\exists x)(\forall y) \sim S[x, y] \qquad (11.15)$$

which is equivalent to

$$\mathrm{K}(\forall x)(\exists y)S[x, y] \qquad (11.12)$$

But (11.12) is weaker than (11.8). What is known according to the former is that everything, say x, is related to something, say y, such that $S[x, y]$. What is known according to the latter is which individual y it is that each given x bears this relation to.

One can perhaps say that indirect proofs can in this way of looking at them yield knowledge that certain mathematical propositions are true, but without knowledge as to why they are true, in the sense that the proof does not show what the Skolem functions are that provide the true-making substitution-values of the quantified propositions in question. The naturalness of the epistemic interpretation of mathematics is illustrated by the fact that formulations of the kind just used have time-honored precedents in the foundational discussion. For instance, one classical formulation of what proofs by contradiction do runs as follows: "Even if they do not give us the cause of why a certain affection is to be predicated of a subject, they nonetheless give us a reason by which we know that a certain state of affairs holds" (Mancosu 1991, p. 34).

Once again, intuitionists have seen something interesting about mathematical reasoning. Once again, their insight turns out to be epistemic in nature. And once again the way intuitionists have tried to spell out their insight is woefully inadequate.

Another application of the epistemic interpretation of logic enables us to interpret certain nonstandard readings of higher-order quantifiers. This interpretation results from taking a second-order sentence S, which we can assume to be in the negation normal form, prefixing it by K, and replacing all second-order qunatifiers like $(\exists X)$ and $(\forall Y)$ by $(\exists X/K)$ and $(\forall Y/K)$, respectively. This idea is readily extended to all higher-order logic. I believe that the resulting interpretation is what some apparent adherents of a nonstandard interpretation of logical and mathematical propositions have had in mind.

Other applications throw light on the idea of intuitionistic logic. In fact, from the vantage point of the epistemic approach to intuitionism we can now see certain serious defects in the attempts that have been made to formulate explicitly an intuitionistic logic, such as Heyting's (1956). By and large, what has happened is the same mistake as was diagnosed above. At best, some unarticulated nonstandard sense of *knowing that* is captured by intuitionistic logic, but not a constructivistic sense of *knowing what*, for example, knowing which function captures a certain regularity. Yet it is the latter issue that is really interesting in the epistemology of mathematics.

For instance, if a formula like

$$(\forall x)(\exists y)S[x, y] \tag{11.7}$$

is supposed to capture the force of

$$K(\forall x)(\exists y/K)S[x, y] \tag{11.8}$$

then universal instantiation will not be a valid rule of logical inference. For (11.8) does not logically imply

$$K(\exists y/K)S[b, y] \tag{11.16}$$

for a given b, as a semantical analysis of the situation will show. Indeed, if we put $S[x, y] = (x = y)$, then (11.8) will say (roughly speaking) that the identity of all individuals is known under some guise or other. From that it does not follow that the identity of the reference of any old nonempty name is known, which is what (11.7) says.

But if (11.8) does not imply (11.16), then (11.7) should not imply

$$(\exists y)S[b, y] \tag{11.17}$$

either, on the intuitionistic interpretation. Yet in intuitionistic logic it does, for universal instantiation is valid in Heyting's logic. Hence intuitionistic logic is seriously inadequate as a true epistemic logic of mathematics. It does not capture the kind of reasoning which is for instance presupposed in intuitionistic and constructivistic criticism of the principle of choice.

This important result calls for a few comments. First, it is independent of the use of explicit epistemic operators. The same change in the rules of quantifier instantiation is needed also in IF first-order logic, as was pointed out in Chapter 3. Hence, it is occasioned by the use of informational independence rather than by the use of epistemic notions. Of course, we do not have a complete set of rules for establishing validity in IF first-order logic, only of rules for establishing inconsistency.

The failure of Heyting's intuitionistic logic admits of an explanation, if not an excuse. The crucial inference from (11.8) to (11.16) can be restored by an additional premise

$$K(\exists x/K)(b = x) \tag{11.18}$$

which says that it is known what (which individual) b is. And of course any old intuitionist would admit that even if he had an effective way of finding a number y, given a number x, such that

$S[x, y]$ is the case, it does not follow that he can do that for b unless he knows which number b is. Hence the failure of Heyting's logic makes perfect sense to an intuitionist.

This intuitionistic meaning of my criticism enables an intuitionist to offer a prima facie reply to my criticism. He can reply that the way his logic is supposed to be applied is different from what I have been assuming. He is willing to substitute for individual constants like the b in (11.17) only such terms whose reference is known to him. This rejoinder nevertheless does not save intuitionistic logic. For one thing, it makes intuitionistic logic incommensurable with ordinary first-order logic. The two are then applied in different ways, for classical logic admits also singular terms whose reference is not known to the logician. Indeed, the use of "arbitrary individuals" is a time-honored ploy in logic. Hence intuitionistic logic is not, on the construal which the intuitionistic response relies on, any longer a rival to classical logic. It is a specialized logic relying on unspoken epistemic assumptions for its applicability.

For another thing, similar failures can be pointed out in intuitionistic logic which are independent of the use of individual constants as substitution-values. Hence, in brief, *Heyting's intuitionistic logic fails its self-imposed task by criteria that intuitionists should accept themselves.*

As was indicated in Chapter 3, this defect in the usual instantiation rules is easily corrected by modifying the instantiation rules which the classicists and the intuitionists have both used. What is needed is a logic in which existential instantiation is effected by means of Skolem functions and in which universal instantiation cannot precede existential instantiation. The requisite changes in instantiation rules are indicated in Chapters 3 and 4. They will do the job here. For instance, (11.7) will now imply

$$(\forall x)S[x, f(x)] \tag{11.19}$$

which is a proxy for

$$K(\forall x)S[x, f(x)] \tag{11.20}$$

Universal instantiation now yields

$$S[b, f(b)] \tag{11.21}$$

which represents

$$KS[b, f(b)] \tag{11.22}$$

As long as the function f is assumed to be known, this yields the right inferential relationships.

The real defect of Heyting's logic does not lie in its choice of inappropriate rules of inference as much as in its failure to incorporate the correct instantiation rules. This failure cannot be corrected by changing its mode of application.

Such modifications in the instantiation rules for quantifiers suffice to restore our disproof rules to a state of semantical completeness. However, it is not obvious that there can exist a semantically complete axiomatization for epistemic logic when it comes to the axiomatization of logical truths. This is a problem even if existential quantifiers and disjunctions are only allowed to be independent of sentence-initial epistemic K-operators but not independent of other quantifiers. An informal explanation of this problem can be given by pointing out that a sentence-initial K is to all effects and model-theoretical purposes a universal quantifier. Hence a sentence of the form

$$K(\exists x)(\forall y)(\exists z/K)S[x, y, z] \tag{11.23}$$

seems to have the same structure as the Henkin prefix,

$$\left.\begin{array}{l} K(\exists x) \\ (\forall y)(\exists z) \end{array}\right\} S[x, y, z] \tag{11.24}$$

And it is known that the logic of Henkin quantifier sentences is not axiomatizable.

Thus Heyting's idea of implementing intuitionistic logic in the customary way, that is, by axiomatizing the logical truths of such a logic, needs a much better justification than it has been given by intuitionists themselves. Intuitionistic logic and mathematics are better viewed model-theoretically than axiomatically.

These observations provide examples of interesting results that hold independently of how functions are thought of as being identified.

One of the virtues of the epistemic interpretation is its generality. The epistemic interpretation allows us to restrict the functions (Skolem functions) which figure in second-order translations of first-order sentences to any class of function that can be considered as "being known". This class could be a class of function wider than the class of recursive ones. Instead of requiring computability, we could merely require of "known" functions that they can be mastered theoretically in some suitable sense. All such variants of the epistemic interpretation can be dealt with by means of a suitable epistemic logic.

However, some possible restrictions could on the contrary be more severe than a restriction to recursive functions. For instance, some-one might want to require some kind of real-time computability.

The observations made in Chapter 4 throw further interesting light on the theoretical situation concerning these different epistemic interpretations. It was seen in Chapter 4 that the criteria of *knowing that* (knowledge of propositions) do not determine the criteria of *knowing what* or *which* (knowledge of objects). This implies that epistemic interpretations of mathematics are in a sense inevitably underdetermined. One mathematician may impose more or less strict requirements on what counts as knowing what given math-ematical functions are, than a colleague of his or hers. And the considerations they can use in order to resolve their disagreement will have to be quite different from the ones used to decide whether a given mathematical proposition is true or not.

From a more general vantage point, the epistemic interpretation of mathematics also offers an interesting general perspective on the epistemology of mathematics. One thing that we might want to do – or at least try out – is to use the epistemic interpretation of mathematics, not as a characterization of what it means for math-ematical propositions to be true, but what it means for them to be known. This possibility is a most natural one, independently of one's stance, in the foundations of mathematics. Even the most Platonistic mathematician might, if only in his or her off-duty hours, raise the question as to in what sense it can be claimed that we *know* the mathematical propositions we are familiar with. Is it really enough to know that in some Platonic heaven there exists verifying strategies for such propositions? Should we also know what those strategies are? Very plausibly, it can be suggested that for the purpose of "really knowing", for instance, the statement

$$(\forall x)(\exists y)S[x, y] \tag{11.7}$$

it is not enough that there exists a choice function f which makes true the formula

$$(\forall x)S[x, f(x)] \tag{11.19}$$

We also have to know an actual instance of a function f which does this.

Thus my epistemic interpretation of mathematics has a great deal of interest for any epistemologist of mathematics, Platonist, constructivist or whatnot. And if so, epistemic logic is already in a position to provide a major insight into the epistemology of mathematics. Without trying to explain the details here, it shows that all questions about whether a certain mathematical proposition S is known, reduce to two simpler ones. They are (i) the question of the truth of S (and/or of the truth of certain related propositions) and (ii) the question as to whether the Skolem functions involved in S are known ones or not.

This provides an extremely interesting perspective on the entire epistemology of mathematics. It shows that the question as to what functions are "known" (in the sense explained above) is not only central in some arcane reinterpretation of logic and mathematics – it is the central question in all epistemology of mathematics.

This observation in turn throws some light on the main lines of the history of mathematics. It shows that a crucial role in the development of mathematical knowledge has been played by the gradual expansion of the notion of function (see here Youschkevitch 1976, Hintikka 1995a). And by expansion I do not mean in the first place the widening of the explicit or implicit definition of an arbitrary function. Rather, what has happened is a broadening of the range of functions that can be said to be known by mathematicians. In other words, what is crucial is the growth of the class of functions that can be mastered intellectually by means of the available mathematical concepts and theories. By choosing the class of known functions in different ways we can thus model the different stages of the historical development of mathematics with the help of my epistemic interpretation of mathematics.

From the vantage point of the epistemic interpretation, a logical axiomatization of a mathematical theory (or even of mathematics at large) in its epistemic dress becomes less of a codification of eternal truths in a permanent fixed language than a report of the current state of mathematical knowledge. For the epistemic impact of the axiom system does not depend only on the truth of the axioms, but also on the class of functions that are known to the users of the axiom system. And this class is not specified by the axiom system itself but only presupposed there. Moreover, it has been seen that such questions of our knowledge of mathematical objects can arise also when there are no explicit function quantifiers in the axiom system in question. All that is needed for the purpose is the presence of dependent existential quantifiers.

As far as the ontology of mathematics is concerned, an epistemic interpretation of mathematics is in any case, in one important respect, on a different footing from the constructivistic one. In the constructivistic one, the range of function variables was restricted to recursive ones. If "to be is to be a value of a bound variable", the constructivistic interpretation in effect denies the existence of nonrecursive functions, at least in so far as strategy functions are concerned.

In contrast, the crucial question in the epistemic interpretation is which functions are known and which ones are not known. Unless one assumes an unrealistically strict positivistic position, this does not presuppose a denial of the existence of functions of other kinds. On the contrary, the way known functions were delineated was by means of function quantifiers ranging over a wider class of arbitrary functions. Thus one can even think of the task of mathematical research as involving an attempt to bring as many of these strange "arbitrary functions" to the fold of intellectually mastered and, in that sense, "known" mathematical objects. This task presupposes, ontologically speaking, existence rather than the inexistence of all and sundry arbitrary functions. In general, it makes sense to speak of mathematical objects as being known or not known and of our coming to know new mathematical objects only if one assumes that they actually exist even when they are not known. A defender of the category of arbitrary functions might thus accuse his or her constructivistic critics of a confusion between epistemological and ontological ideas. In the last analysis, logical rather than Freudian, the epistemic interpretation on mathematics thus aids and abets the classical view of mathematical existence, as long as we realize that this view has to be complemented by epistemic considerations.

The same point can be put in different terms. Once the epistemic element in mathematics is made explicit, the intuitionistic criticisms of the classical sense of *knowing that* and hence of classical mathematics lose their sting. This applies also to the questions whether certain kinds of mathematical objects in fact exist or not. Indeed, an approach to the epistemology of mathematical objects (such as we found the genuine Brouwerian intuitionist undertaking to be) scarcely makes sense unless the unqualified existence of mathematical objects is taken in the classical sense. Thus the ultimate thrust of the intuitionists' philosophy, rightly understood, supports rather than undermines the kind of classical view of the ontology of mathematics that goes together with the standard interpretation of the existence of higher-order entities. Intuitionists are in the last analysis closet realists.

This point is obscured in the literature on the foundations of mathematics by a failure to distinguish from each other the question as to which mathematical objects exist and the question as to which ones are known. This confusion is not a recent phenomenon. On the contrary, realizing that such a confusion has actually affected mathematicians helps us to understand what otherwise might appear to be contradictory statements by one and the same mathematician. A representative example is offered by no lesser a figure than Leonhard Euler. On the one hand, he has been hailed not merely as a precursor but even as a full-fledged believer in arbitrary functions. And we do in fact find statements like the following:

> If some quantities so depend on other quantities that as the latter are changed the former undergo change, then the former quantities are called functions of the latter. This denomination is of the broadest nature and comprises every method by means of which one quantity could be determined by others. If, therefore, x denotes a variable quantity, then all quantifiers which depend upon x in any way or are determined by it are called functions of it.

> (*Institutiones calculi differentialis*, quoted after Bottazzini 1986, p. 33.)

This sounds conclusive (that is, general) enough. But in Euler (1988) we read:

> A function of a variable quantity is an analytic expression composed in any way whatsoever of the variable quantity and numbers or constant quantities.

The obvious explanation is that Euler thought only of functions represented by analytic expressions as being known to him. So why did he bother to preserve the possibility of a wider conception of a function? Once again the actual facts of the history of mathematics yield an interesting clue. His work on the problem of the vibrating string had led him to the conclusion that there may in fact be modes of dependence in physical nature that are discontinuous and hence cannot be captured by analytic expressions (see Bottazzini 1986, pp. 21–33). In other words, problems of applied mathematics made him aware that there may very well exist functions that are not yet known, even though Euler did not express himself in this way.

This example illustrates the fact that the epistemic approach to the foundations of mathematics is very closely related to the realities of the actual historical development of mathematics.

APPENDIX

IF First-Order Logic, Kripke, and 3-Valued Logic

GABRIEL SANDU

In this appendix we show how the result of the definability of truth in IF first-order logic in the sense discussed in Chapter 6 relates to Kripke's result of the definability of truth in partial models. For this purpose, we will give a short description of partial models together with Kripke's result. Before doing that, we shall have to recast the description of IF first-order logic in a form which makes comparison with partial models possible. (The notation used here should be self-explanatory even though it differs slightly from the one used in the bulk of this book.)

1. IF first-order logic

Let us fix an IF first-order language L in an arbitrary signature. As we have seen, this language is nothing else but an ordinary first-order language enriched with a slash "/". In this new language we shall have formulas like

$$\forall x \exists y \forall z (\exists w / \forall x) \phi \tag{A.1.1}$$

or

$$\forall x \exists y \forall z (\phi \, (\vee / \forall x) \, \psi) \tag{A.1.2}$$

We shall denote formulas like (A.1.1) by $G_{1;1}^{1;1} xyzw\phi$, and formulas like (A.1.2) by $V_{1;2}^{1;1} xyz(\phi, \psi)$. More generally, we shall adopt the

254

following conventions:

$G_{k,l}^{n,m} x_1 \ldots x_n y_1 \ldots y_m z_1 \ldots z_k w_1 \ldots w_l \phi$ denotes

$\forall x_1 \ldots x_n \exists y_1 \ldots y_m \forall z_1 \ldots z_k (\exists w_1 / \forall x_1 \ldots x_n) \ldots (\exists w_l / \forall x_1 \ldots x_n) \phi$

$V_{k,2}^{n,m} x_1 \ldots x_n y_1 \ldots y_m z_1 \ldots z_k (\phi, \psi)$ denotes

$\forall x_1 \ldots x_n \exists y_1 \ldots y_m \forall z_1 \ldots z_k (\phi (\vee / \forall x_l \ldots x_n) \psi)$

$G_{k,1}^{*n,m} x_1 \ldots x_n y_1 \ldots y_m z_1 \ldots z_k w_1 \ldots w_l \phi$ denotes

$\exists x_1 \ldots x_n \forall y_1 \ldots y_m \exists z_1 \ldots z_k (\forall w_1 / \exists x_1 \ldots x_n) \ldots (\forall w_l / \forall x_1 \ldots x_n) \phi$

$V_{k,2}^{*n,m} x_1 \ldots x_n y_1 \ldots y_m z_1 \ldots z_k (\phi, \psi)$ denotes

$\exists x_1 \ldots x_n \forall y_1 \ldots y_m \exists z_1 \ldots z_k (\phi (\wedge / \exists x_1 \ldots x_n) \psi)$ ∎

We let the formulas of L be the smallest class of formulas containing the usual atomic formulas and closed under $\neg, \vee, \wedge, \exists, \forall, G_{k,l}^{n,m}, G_{k,l}^{*n,m}, V_{k,2}^{n,m},$ and $V_{k,2}^{*n,m}(n, m, k, l \geqslant 1)$. Notice that $G_{1,0}^{1,0} x \phi = G_{1,0}^{0,0} x \phi = \forall x \phi$, $V_{0,2}^{0,0}(\phi, \psi) = (\phi \wedge \psi)$, $V_{0,2}^{*0,0}(\phi, \psi) = (\phi \wedge \psi)$.

With every formula $\phi(\bar{x})(\bar{x} = x_1, \ldots, x_n)$ of L and model M of L, a game $G(\phi(\bar{x}), M, g)$ is associated, where g is an assignment restricted to the free variables $\bar{x}$ of ϕ with values in $\text{dom}(M)$. If $g(x_1) = a_1, \ldots, g(x_n) = a_n$, we prefer to use the notation $G(\phi(\bar{x}), (M, a_1, \ldots, a_n))$ instead of $G(\phi(\bar{x}), M, g)$. The game is played by two players: Myself (the initial verifier) and Nature (the initial falsifier). The definition of $G(\phi(\bar{x}), M, a_1, \ldots, a_n)$ is by induction on the complexity of ϕ:

Definition 1.

(i) $G(\phi(\bar{x}), M, g)$, for $\phi(\bar{x})$ atomic, contains no moves. If $(M, g) \models \phi(\bar{x})$, then the player who is the verifier wins $G(\phi(\bar{x}), M, g)$; otherwise, the player who is the falsifier wins.

(ii) $G(\neg \phi, M, g)$ is identical with $G(\phi, M, g)$, except that the players exchange roles.

(iii) $G((\phi_0 \vee \phi_1), M, g)$ starts by the verifier choosing $i \in \{0, 1\}$. The game goes on as $G(\phi_i, M, g)$.

(iv) $G((\phi_0 \wedge \phi_1), M, g)$ starts by the falsifier choosing $i \in \{0, 1\}$. The game goes on as $G(\phi_i, M, g)$.

(v) $G(\exists x \phi, M, g)$ starts with the verifier choosing an element $a \in \text{dom}(M)$. The game goes on as in $G(\phi, M, g \cup \{(x, a)\})$.

(vi) $G(\forall x \phi, M, g)$ starts with the falsifier choosing an element $a \in \text{dom}(M)$. The game goes on as in $G(\phi, M, g \cup \{(x, a)\})$.

(vii) $G(G_{k,1}^{n,m}x_1 \ldots x_n y_1 \ldots y_m z_1 \ldots z_k w_1 \ldots w_l \phi, M, g)$ is played in the following way: the falsifier chooses a sequence $\bar{a}$ of length n, then the verifier chooses a sequence $\bar{b}$ of length m, after which the verifier chooses a sequence $\bar{c}$ of length k, and then finally the verifier chooses $\bar{d} = (d_1, \ldots, d_1)$. The game goes on as in $G(\phi, M, g \cup \{(\bar{x}, \bar{a})\} \cup \{(\bar{y}, \bar{b})\} \cup \{(\bar{z}, \bar{c})\} \cup \{(\bar{w}, \bar{d})\})$.

(viii) $G(G_{k,l}^{*n,m}x_1 \ldots x_n y_1 \ldots y_m z_1 \ldots z_k w_1 \ldots w_l \phi, M, g)$ is played exactly as $G(G_{k,1}^{n,m}x_1 \ldots x_n y_1 \ldots y_m z_1 \ldots z_k w_1 \ldots w_l \phi, M, g)$, except that the moves made by the verifier in the former are made by the falsifier in the latter.

(ix) $G(V_{k,2}^{n,m}x_1 \ldots x_n y_1 \ldots y_m z_1 \ldots z_k (\phi_0, \phi_1), M, g)$ is played analogously to $G(G_{k,l}^{n,n}x_1 \ldots x_n y_1 \ldots y_m z_1 \ldots z_k w_1 \ldots w_l \phi, M, g)$, except that in the last move the verifier chooses $i \in \{0, 1\}$, and the game goes on as in $G(\phi_i, M, g \cup \{(\bar{x}, \bar{a})\} \cup \{(\bar{y}, \bar{b})\} \cup \{(\bar{z}, \bar{c})\})$.

(x) The description of the game $G(V_{k,2}^{*n,m}x_1 \ldots x_n y_1 \ldots y_m z_1 \ldots z_k (\phi_0, \phi_1), M, g)$ should be obvious by now.

The notion of strategy is crucial here. A strategy for a player p (p is either Myself or Nature) in the game $G(\phi, M, g)$ is a set $F_p(G(\phi, M, g)) = \{f_1, \ldots, f_n\}$ of functions $f_1, \ldots, f_n$ each f_i corresponding to a move M_i of the player p in the game. Usually, each f_i is defined on the set of all the possible moves of the opponent of p before the move M_i, with the following exceptions: in the game $G(G_{k,l}^{n,m}x_1 \ldots x_n y_1 \ldots y_m z_1 \ldots z_k w_1 \ldots w_l \phi, M, g)$, the function f_i associated with the move of the verifier prompted by w_i is not defined on $(\text{dom}(M))^{n+k}$, but on $(\text{dom}(M))^k$. In other words this game is one of imperfect information. Similarly for the game $G(V_{k,2}^{n,m}x_1 \ldots x_n y_1 \ldots y_m z_1 \ldots z_k (\phi_0, \phi_1), M, g)$. Of course the same applies to the moves of the falsifier in $G(G_{k,l}^{*n,m}x_1 \ldots x_n y_1 \ldots y_m z_1 \ldots z_k w_1 \ldots w_l \phi_0, M, g)$, and in $G(V_{k,2}^{*n,m}x_1 \ldots x_n y_1 \ldots y_m z_1 \ldots z_k (\phi_0, \phi_1), M, g)$.

A strategy for a player in the game $G(\phi, M, g)$ is a winning one if the player wins $G(\phi, M, g)$ no matter what the moves of the opponent are.

The following proposition is straightforward from the definitions:

Proposition 1. For any formula ϕ, model M of L, and assignment g in M (restricted to the free variables of ϕ) it is not possible that both Myself and Nature have a winning strategy in $G(\phi, M, g)$.

In Chapter 2 the truth of a sentence ϕ in the model M was defined as the existence of a winning strategy for Myself in the game $G(\phi, M)$.

It is convenient, for later comparisons, to have a notation for both truth and falsity in M. For this purpose, whenever ϕ is a formula, we use $M \vDash_{\text{GTS}} \phi^+$ to mean 'ϕ is true in M', and $M \vDash_{\text{GTS}} \phi^-$ to mean 'ϕ is false in M'.

Definition 2. Let ϕ be a formula of L, M a model of L, and g an assignment in M restricted to the free variables of ϕ. Then

(i) $(M, g) \vDash_{\text{GTS}} \phi^+$ iff Myself has a winning strategy in $G(\phi, M, g)$.
(ii) $(M, g) \vDash_{\text{GTS}} \phi^-$ iff Nature has a winning strategy in $G(\phi, M, g)$.

We shall now define by double induction two mappings * and #, the truth-preserving and the falsity-preserving mapping, respectively, which map formulas of L into formulas of L:

Definition 3. The mappings * and # are defined as follows:

(i) $\phi^* = \phi$; $\phi\# = \neg\phi$, for ϕ atomic.
(ii) $(\neg\phi)^* = (\phi\#$; $(\neg\phi)\# = \phi^*$.
(iii) $(\phi \vee \psi)^* = (\phi^* \vee \psi^*)$; $(\phi \vee \psi)\# = (\phi\# \wedge \psi\#)$.
(iv) $(\phi \wedge \psi)^* = (\phi^* \wedge \psi^*)$; $(\phi \wedge \psi)\# = (\phi\# \vee \psi\#)$.
(v) $(\exists x\phi)^* = \exists x\phi^*$; $(\exists x\phi)\# = \forall x\phi\#$.
(vi) $(\forall x\phi)^* = \forall x\phi^*$
$\quad (\forall x\phi)\# = \exists x\phi\#$
(vii) $G_{k,l}^{n,m} x_1 \ldots x_n y_1 \ldots y_m z_1 \ldots z_k w_1 \ldots w_1 \phi)^*$
$\quad = G_{k,l}^{n,m} x_1 \ldots x_n y_1 \ldots y_m z_1 \ldots z_k w_1 \ldots w_1 \phi^*$
$\quad (G_{k,l}^{n,m} x_1 \ldots x_n y_1 \ldots y_m z_1 \ldots z_k w_1 \ldots w_1)\#$
$\quad = G_{k,l}^{*n,m} x_1 \ldots x_n y_1 \ldots y_m z_1 \ldots z_k w_1 \ldots w_l \phi\#.$
(viii) $G_{k,l}^{*n,m} x_1 \ldots x_n y_1 \ldots y_m z_1 \ldots z_k w_1 \ldots w_l \phi)^*$
$\quad = G_{k,l}^{*n,m} x_1 \ldots x_n y_1 \ldots y_m z_1 \ldots z_k w_1 \ldots w_l \phi)^*$;
$\quad (G_{k,l}^{*n,m} x_1 \ldots x_n y_1 \ldots y_m z_1 \ldots z_k w_1 \ldots w_l \phi)\#$
$\quad = G_{k,l}^{n,m} x_1 \ldots x_n y_1 \ldots y_m z_1 \ldots z_k w_1 \ldots w_l \phi\#.$
(ix) $(V_{k,2}^{n,m} x_1 \ldots x_n y_1 \ldots y_m z_1 \ldots z_k (\phi_0, \phi_1))^*$
$\quad = V_{k,2}^{n,m} x_1 \ldots x_n y_1 \ldots y_m z_1 \ldots z_k (\phi_0^*, \phi_1^*))$;
$\quad (V_{k,2}^{n,m} x_1 \ldots x_n y_1 \ldots y_m z_1 \ldots z_k (\phi_0, \phi_1))\#$
$\quad = V_{k,2}^{*n,m} x_1 \ldots x_n y_1 \ldots y_m z_1 \ldots z_k (\phi_0^\#, \phi_1^\#).$
(x) $(V_{k,2}^{*n,m} x_1 \ldots x_n y_1 \ldots y_m z_1 \ldots z_k (\phi_0, \phi_1))^*$
$\quad = V_{k,2}^{*n,m} x_1 \ldots x_n y_1 \ldots y_m z_1 \ldots z_k (\phi_0^*, \phi_1^*)$;
$\quad (V_{k,2}^{*n,m} x_1 \ldots x_n y_1 \ldots y_m z_1 \ldots z_k (\phi_0, \phi_1))\#$
$\quad = V_{k,2}^{n,m} x_1 \ldots x_n y_1 \ldots y_m z_1 \ldots z_k (\phi_0^\#, \phi_1^\#).$

A straightforward induction on the complexity of ϕ shows that

Proposition 4. For any formula ϕ, model M of L, and assignment g:

(i) Myself hs a winning strategy in $G(\phi, M, g)$ iff Myself has a winning strategy in $G(\phi^*, M, g)$.

(ii) Naure has a winning strategy in $G(\phi, M, g)$ iff Myself has a winning strategy in $G(\phi\#, M, g)$.

Corollary. For any formula ϕ:

(i) $(M, g) \vDash_{\text{GTS}} \phi^+$ iff $(M, g) \vDash_{\text{GTS}} (\phi^*)^+$

(ii) $(M, g) \vDash_{\text{GTS}} \phi^-$ iff $(M, g) \vDash_{\text{GTS}} (\phi\#)^+$

Proof: from Proposition 4 and Definition 1.

Proposition 5. Let ϕ be an arbitrary IF first-order sentence. Then the following holds for any model M:

(i) $M \vDash_{\text{GTS}} (\neg \phi)^+$ iff $M \vDash_{\text{GTS}} \phi^-$

(ii) $M \vDash_{\text{GTS}} (\neg \phi)^-$ iff $M \vDash_{\text{GTS}} \phi^+$

(iii) $M \vDash_{\text{GTS}} (\phi \vee \psi)^+$ iff $M \vDash_{\text{GTS}} \phi^+$ or $M \vDash_{\text{GTS}} \psi^+$

(iv) $M \vDash_{\text{GTS}} (\phi \vee \psi)^-$ iff $M \vDash_{\text{GTS}} \phi^-$ and $M \vDash_{\text{GTS}} \psi^-$

(v) $M \vDash_{\text{GTS}} (\phi \wedge \psi)^+$ iff $M \vDash_{\text{GTS}} \phi^+$ and $M \vDash_{\text{GTS}} \psi^+$

(vi) $M \vDash_{\text{GTS}} (\phi \wedge \psi)^-$ iff $M \vDash_{\text{GTS}} \phi^-$ or $M \vDash_{\text{GTS}} \psi^-$

Proof: (i) From the Corollary and Definition 3 we have $M \vDash_{\text{GTS}} (\neg \phi)^+$ iff $M \vDash_{\text{GTS}} ((\neg \phi)^*)^+$ iff $M \vDash_{\text{GTS}} (\phi\#)^+$ iff $M \vDash_{\text{GTS}} \phi^-$. All the other cases are similar. ∎

We know that the law of excluded middle fails in IF first-order logic. This raises the question – what is the valuation schema that is satisfied by the Boolean connectives in this logic.

In order to give an answer to this question, let $\|\phi\|^M$ denote the truth-value of the L-sentence ϕ in the model M. Putting

$\|\phi\|^M = 1$, if $M \vDash_{\text{GTS}} \phi^+$ (Hence, by Proposition 1: not $M \vDash_{\text{GTS}} \phi^-$)

$\|\phi\|^M = 0$, if $M \vDash_{\text{GTS}} \phi^-$ (Hence, by Proposition 1: not $M \vDash_{\text{GTS}} \phi^+$)

$\|\phi\|^M = ?$ if not $M \vDash_{\text{GTS}} \phi^+$ and not $M \vDash_{\text{GTS}} \phi^-$

it is easy to check, using Proposition 5, that the Boolean connectives satisfy the Strong Kleene valuation schema (conjunction, and implication are defined via negation and disjunction in the usual way):

ϕ	ψ	$\neg\phi$	$\neg\psi$	$(\phi\vee\psi)$	$(\phi\wedge\psi)$	$(\phi\rightarrow\psi)$	$(\phi\leftrightarrow\psi)$
1	1	0	0	1	1	1	1
1	0	0	1	1	0	0	0
1	?	0	?	1	?	?	?
0	1	1	0	1	0	1	0
0	0	1	1	0	0	1	1
0	?	1	?	?	0	1	?
?	1	?	0	1	?	1	?
?	0	?	1	?	0	?	?
?	?	?	?	?	?	?	?

It will be seen below that the partial logic used by Kripke in his result about the definability of truth satisfies the Strong Kleene valuation schema too.

2. IF first-order logic and the definability of truth and falsity

In this section we shall resume the argument of Chapter 6 showing that an IF first-order language (in the signature of Peano Arithmetic) admits of a truth-predicate in this language itself.

So let the signature S of the language L be that of Peano Arithmetic (i.e., $S = \{0, 1, +, \cdot, \text{and} <\}$).

A couple of remarks are in order here.

Any formula of the form $V^{n,m}_{k,2}x_1\ldots x_ny_1\ldots y_mz_1\ldots z_k(\phi_0, \phi_1)$ is definable from $G^{n,m}_{k,l}x_1\ldots x_ny_1\ldots y_mz_1\ldots z_kw_1\ldots w_l\phi$ and the standard quantifiers:

$$V^{n,m}_{k,2}x_1\ldots x_ny_1\ldots y_mz_1\ldots z_k(\phi_0, \phi_1) \Leftrightarrow$$
$$G^{n,m}_{k,1}x_1\ldots x_ny_1\ldots y_mz_1\ldots z_kw_1((w_1 = 0 \wedge \phi_0) \vee (w_1 = 1 \wedge \phi_1))$$

(A.2.1)

So in the sequel we shall disregard formulas $V^{n,m}_{k,2}x_1\ldots x_ny_1\ldots y_mz_1\ldots z_k(\phi_0, \phi_1)$.

Also, we shall assume that the formulas of L are closed only with respect to $G^{1,1}_{1,1}x_1y_1z_1w_1\phi$, instead of being closed with respect to every $G^{n,m}_{k,l}x_1\ldots x_ny_1\ldots y_mz_1\ldots z_kw_1\ldots w_l\phi$. (cf. our remarks below).

Let us fix a Gödel numbering of L. We let ⌈e⌉ be the Gödel number of the expression e. We shall use the following recursive functions and relations (in addition to *, and #):

ct(n): n is a closed term of L

v(n): the value of n, if n is a closed term of L

fml(n): n is an IF first-order formula of L

st(n): n is an IF first-order sentence of L

Sub(m, n, p): the substitution function whose value is $\ulcorner\phi(p)\urcorner$, if m is the Gödel number of the formula $\phi(x)$, and n is the Gödel number of the free variable x; and m otherwise (p is the numeral corresponding to p).

∎

A Σ_1^1-formula over L is a formula having the form $\exists f_1^{r_1}\ldots f_n^{r_n}\psi$, where ψ is a first-order formula (in the vocabulary of L plus $\{f_1^{r_1}, \ldots, f_n^{r_n}\}$).

Let $L(X)$ (X is a second-order variable) be the conjunction of the sentences

$$\forall x(\mathrm{st}(x) \to (X(x) \leftrightarrow X(x^*))) \tag{A.2.2}$$

$$\forall x, y((\mathrm{ct}(x) \wedge \mathrm{ct}(y)) \to (X(\ulcorner(x = y)\urcorner) \leftrightarrow v(x)$$
$$= v(y)) \wedge (X(\ulcorner(x < y)\urcorner) \leftrightarrow v(x) < v(y)) \wedge (X(\ulcorner\neg(x < y)\urcorner) \leftrightarrow \neg v(x)$$
$$< v(y)) \wedge (X(\ulcorner\neg(x = y)\urcorner) \leftrightarrow \neg v(x) = v(y))) \tag{A.2.3}$$

$$\forall x, y((\mathrm{st}(x) \wedge \mathrm{st}(y)) \to (X(\ulcorner(x \wedge y)\urcorner) \to X(x) \wedge X(y))) \tag{A.2.4}$$

$$\forall x, y((\mathrm{st}(x) \wedge \mathrm{st}(y)) \to (X(\ulcorner(x \vee y)\urcorner) \to X(x) \vee X(y))) \tag{A.2.5}$$

$$\forall y((fl(y) \wedge \mathrm{st}(\ulcorner\forall x_i y\urcorner)) \to (X(\ulcorner(\forall x_i y\urcorner) \to \forall x X(\mathrm{Sub}(y, \ulcorner x_i\urcorner, x))))) \tag{A.2.6}$$

$$\forall y((fl(y) \wedge \mathrm{st}(\ulcorner\exists x_i y\urcorner)) \to (X(\ulcorner(\exists x_i y\urcorner) \to \exists x X(\mathrm{Sub}(y, \ulcorner x_i\urcorner, x))))) \tag{A.2.7}$$

$$\forall y((fl(y) \wedge \mathrm{st}(\ulcorner G_{1,1}^{1,1} x_1 y_1 z_1 w_1 y\urcorner)) \to (X(\ulcorner G_{1,1}^{1,1} x_1 y_1 z_1 w_1 y\urcorner) \to$$
$$\exists fg \forall x_1 z_1 X(\mathrm{Sub}(\mathrm{Sub}(\mathrm{Sub}(\mathrm{Sub}(y, \ulcorner y_1\urcorner, f(x_1)), \ulcorner w_1\urcorner, g(z_1)),$$
$$\ulcorner x_1\urcorner, t), \ulcorner z_1\urcorner, z_1)))) \tag{A.2.8}$$

$$\forall y((fl(y) \wedge \mathrm{st}(\ulcorner G_{1,1}^{*1,1} x_1 y_1 z_1 w_1 y\urcorner)) \to (X(\ulcorner G_{1,1}^{*1,1} x_1 y_1 z_1 w_1 y\urcorner) \to$$
$$\exists f x_1 \forall y_1 w_1 X(\mathrm{Sub}(\mathrm{Sub}(\mathrm{Sub}(\mathrm{Sub}(y, \ulcorner x_1\urcorner, x_1)), \ulcorner z_1\urcorner, f(y_1)),$$
$$\ulcorner y_1\urcorner, y_1), \ulcorner w_1\urcorner, w_1))) \tag{A.2.9}$$

∎

Remark. The expressions '$\neg$', '$\wedge$', '$\vee$', '$\forall$', '$\exists$', '$G_{1,1}^{1,1}$', and '$G_{1,1}^{*1,1}$' occurring within the corners $\ulcorner\urcorner$ do not have their usual meaning but are function symbols which name primitive recursive functions. Thus

'$\neg$' denotes a function f defined by: $f(n) = \ulcorner \neg \phi \urcorner$, if n is the Gödel number of the formula ϕ; and n otherwise. '$\forall$' denotes a binary function, g defined by: $g(n, m) = \ulcorner \forall x\phi \urcorner$, if n is the Gödel number of the formula ϕ and m is the Gödel number of the variable x; and n otherwise. Similarly, '$G^{1;1}_{1;1}$' denotes a function h defined by: $h(n, m, p, q, r) = \ulcorner G^{1;1}_{1;1} x_1 y_1 z_1 w_1 \phi \urcorner$, if n is the Gödel number of the formula ϕ, m is the Gödel number of the variable x_1, p is the Gödel number of the variable y_1, q is the Gödel number of the variable z_1, and m is the Gödel number of the variable w_1; and n otherwise. The other cases should be obvious by now.

Proposition 1. Let $\theta(x)$ be the Σ^1_1-formula $\exists X(L(X) \wedge X(x))$, and N the standard model of L. Then for every IF first-order formula ϕ

$$N \vDash \theta(\ulcorner \phi \urcorner) \text{ iff } N \vDash_{\text{GTS}} \phi^+ \tag{A.2.10}$$

(In other words, $\theta(x)$ defines explicitly 'truth in N' for the language L.)

Proof of (A.2.10): Assume $N \vDash_{\text{GTS}} \phi^+$, and, in order to prove $N \vDash (\exists X)(L(X) \wedge X(\ulcorner \phi \urcorner))$, let $A = \{\ulcorner \phi \urcorner: \phi \text{ is a sentence of } L \text{ and } N \vDash_{\text{GTS}} \phi^+\}$. An induction on the complexity of ϕ shows that $(N, A) \vDash L(X) \wedge X(\ulcorner \phi \urcorner)$. In the induction, it is enough to consider only the cases for ϕ being in negation normal form, since ϕ is in X only if ϕ^* is in X, and ϕ^* is in negation normal form. Let us prove two cases.

Assume ϕ is $\forall x\psi$ and $N \vDash_{\text{GTS}} (\forall x\psi)^+$. The inductive hypothesis is

$$\text{If } N \vDash_{\text{GTS}} \chi^+ \text{ then } (N, A) \vDash L(X) \wedge X(\ulcorner \chi \urcorner) \tag{+}$$

for every sentence χ of complexity less than that of $\forall x\psi$.

From $N \vDash_{\text{GTS}} (\forall x\psi)^+$, it follows that $A(\ulcorner \forall x\psi \urcorner)$. Also, since Myself has a winning strategy in $G(\forall x\psi, N)$, Myself has to win against every move of Nature, that is, Myself has a winning strategy in every game $G(\psi(x), N, n)$, for every $n \in N$. Hence $(N, n) \vDash_{\text{GTS}} \psi(x)^+$, for every n, or equivalently, $N \vDash_{\text{GTS}} \psi(n)^+$, for every n. Hence, by $(+)$, $A(\ulcorner \psi(n) \urcorner)$, for every $n \in N$, that is, $(N, A) \vDash \forall x X(\text{Sub}(\ulcorner \psi \urcorner, x))$.

Assume now ϕ is $G^{1;1}_{1;1} x_1 y_1 z_1 w_1 \psi$, and $N \vDash_{\text{GTS}}(G^{1;1}_{1;1} x_1 y_1 z_1 w_1 \psi (x_1 y_1 z_1 w_1))^+$. Then for every m picked up by Nature, Myself finds $p = f(m)$, and for every n picked up by Nature, Myself finds $s = g(n)$, (f, and g are part of the winning strategy of Myself in the game) such that Myself wins the game $G(\psi(x_1, y_1, z_1, w_1), N, m, p, n, s)$, that is, $(N, m, p, n, s) \vDash_{\text{GTS}} (\psi(x_1, y_1, z_1 w_1))^+$. Equivalently, $N \vDash_{\text{GTS}} (\psi(m, p, n, s))^+$.

By the inductive hypothesis, $A(^{\ulcorner}\psi(m,\ p,\ n,\ s)^{\urcorner})$, for every quadruple $m,\ p, n, s \in N$, such that $p = f(m)$, and $s = g(n)$. Hence $(N, A) \vDash \exists fg \forall x_1 z_1 X(\mathrm{Sub}(\mathrm{Sub}(\mathrm{Sub}(\mathrm{Sub}(y, {}^{\ulcorner}y^{\urcorner}, f(x_1)), {}^{\ulcorner}w_1^{\urcorner}, g(z_1)), {}^{\ulcorner}x_1^{\urcorner}, x_1), {}^{\ulcorner}z_1^{\urcorner}, z_1))))$.

The other cases are similar.

In the other direction, assume $N \vDash (\exists X)(L(X) \wedge X(^{\ulcorner}\phi^{\urcorner}))$. We prove by induction on the complexity of ϕ that $N \vDash_{\mathrm{GTS}} \phi^+$. The only complicated cases are for ϕ being $\forall x \psi$, and $G_{1,1}^{1,1} x_1 y_1 z_1 w_1 \psi$.

Suppose ϕ is $\forall x \psi$. By our assumption, there is a set A such that

$$(N, A) \vDash L(X) \wedge X(^{\ulcorner}\forall x \psi^{\urcorner}) \tag{A.2.11}$$

which implies

$$(N, A) \vDash L(X) \wedge X(^{\ulcorner}\psi(n)^{\urcorner}) \tag{A.2.12}$$

for all $n \in N$. Hence

$$(N, \vDash (\exists X)(L(X) \wedge X(^{\ulcorner}\psi(n)^{\urcorner})) \tag{A.2.13}$$

for all $n \in N$. By the inductive hypothesis

$$N \vDash_{\mathrm{GTS}} \psi(n)^+ \tag{A.2.14}$$

for all $n \in N$, that is, $(N, n) \vDash_{\mathrm{GTS}} \psi(x)^+$, for all $n \in N$. Thus Myself has a winning strategy in every game $G(\psi(x), N, n)$, for arbitrary n. Then it is trivial to see that Myself has a winning strategy in $G(\forall x \psi, N)$. For let Nature choose an individual $n \in N$ and the game go on with $G(\psi(x), N, n)$. Then Myself does not need to do anything else than follow his winning strategy in $G(\psi(x), N, n)$.

Suppose now that ϕ is $G_{1,1}^{1,1} x_1 y_1 z_1 w_1 \psi(x_1 y_1 z_1 w_1)$ and that

$$N \vDash (\exists X)(L(X) \wedge X(^{\ulcorner}G_{1,1}^{1,1} x_1 y_1 z_1 w_1 \psi(x_1 y_1 z_1 w_1)^{\urcorner})) \tag{A.2.15}$$

In a similar way to the previous case, we get that

$$(N, A) \vDash L(X) \wedge \exists fg \forall x_1 z_1 X(\mathrm{Sub}(\mathrm{Sub}(\mathrm{Sub}(\mathrm{Sub}({}^{\ulcorner}\psi^{\urcorner}, {}^{\ulcorner}y_1^{\urcorner}, f(x_1)),$$

$$ {}^{\ulcorner}w_1^{\urcorner}, g(z_1)), {}^{\ulcorner}x_1^{\urcorner}, x_1), {}^{\ulcorner}z_1^{\urcorner}, z_1)))) \tag{A.2.16}$$

Hence

$$(N, A) \vDash L(X) \wedge X(^{\ulcorner}\psi(m,\ p,\ n,\ s^{\urcorner}) \tag{A.2.17}$$

for every quadruple $(m,\ p,\ n,\ s)$ which is such that $p = f(m)$, and $s = g(n)$, for some functions f and g. As above, by the inductive hypothesis we have

$$N \vDash_{\text{GTS}} \psi(m, p, n, s)^+ \text{ (equivalently } (N, m, p, n, s)$$

$$\vDash_{\text{GTS}} \psi(x_1, y_1, z_1, w_1)^+) \tag{A.2.18}$$

for every quadruple (m, p, n, s) such that $p = f(m)$, and $s = g(n)$. Again, it is easy to see that Myself has a winning strategy in $G(G_{1;1}^{1;1} x_1 y_1 z_1 z_1 \psi, N)$. For let Nature choose arbitrary m; then Myself chooses p such that $p = f(m)$. Then let Nature choose again, n; Myself chooses s such that $s = g(n)$, and the game goes on with $G(\psi(x_1, y_1, z_1, w_l, N, m, p, n, s)$. In this game, let Myself follow his winning strategy which exists by (A.2.17).

The formula $\theta(x)$ which defines 'truth in N' for the IF first-order language L (strictly speaking, for the subfragment of L closed with respect to the standard quantifiers, the Boolean connectives and $G_{1;1}^{1;1} x_1 y_1 z_1 w_1$) is a Σ_1^1-formula. However, it was pointed out in Chapter 6 that each such formula is equivalent to an IF first-order formula. To show that $\theta(x)$ is equivalent to a formula of the relevant IF first-order subfragment (i.e., the subfragment closed with respect to the standard quantifiers, the Boolean connectives and $G_{1;1}^{1;1} x_1 y_1 z_1 w_1$) one needs the following further observations.

In Walkoe (1970) an effective procedure is given which translates every Σ_1^1-formula ϕ of an arbitrary language L into a formula λ_ϕ which is logically equivalent to ϕ, such that λ_ϕ belongs to a language L^* which extends ordinary first-order logic with all Henkin quantifiers (more exactly, there is an additional clause saying that, if ϕ is a first-order formula, then $Q\phi$ is a formula of L^*, where Q is defined below).

A Henkin quantifier Q is usually defined as a triple $Q = (A_Q, E_Q, D_Q)$, where A_Q and E_Q are disjoint sets of variables (universal and existential variables, respectively) and D_Q (the dependency relation) is a partial ordering on $A_Q \times E_Q$. Intuitively, if $(x, y) \in D_Q$ then the existentially quantified variable y is in the scope of the universally quantified variable x. For instance, the Henkin quantifier $Q = (\{x, z\}, \{y, w\}, \{(x, y), (z, w)\})$ says that the quantifier $\exists y$ is in the scope of $\forall x$, and that $\exists w$ is in the scope of $\forall z$. The interpretation of a Henkin quantifier is given through a second-order translation. For instance, the interpretation of Q is given by

$$Qxyzw\phi(x, y, z, w) \Leftrightarrow \exists f \exists g \forall x \forall z \phi(x, f(x), z, g(z)) \tag{A.2.19}$$

In Krynicki (1993) it is shown that in models with definable pairing functions every formula of L^* is materially equivalent to a formula of the subfragment of L^* in which the only Henkin quantifier is Q (from A.2.19). On the other side, Q is definable in IF first-order language, that is

$$Qxyzw\phi(x, y, z, w) \Leftrightarrow G_{1,1}^{1,1} xyzw\phi(x, y, z, w) \qquad (A.2.20)$$

Thus, since N has definable pairing functions, we have shown that the formula $\theta(x)$ is equivalent in N with a formula of the IF first-order language L. ∎

3. Partial models

The easiest way to introduce partial models is as models of a propositional language L consisting of a set P of primitive propositional symbols $S_1, S_2, \ldots$. The sentences of L are built up from P using negation and disjunction in the usual way. A partial model M of L is a triple $M = (A, V^+, V^-)$, where A is a subset of P and V^+ and V^- are disjoint subsets of M. The idea is that V^+ represents the true sentences of L, and V^- the false ones. For S a sentence of L we use $M \vDash_{PM} S^+$ to mean 'S is true in A', and $A \vDash_{PM} S^-$ to mean 'S is false in A'. Truth and falsity of a sentence S in A are defined by induction in the following way:

$$M \vDash_{PM} S^+ \text{ iff } S \in V^+, S \text{ primitive}$$

$$M \vDash_{PM} S^- \text{ iff } S \in V^-, S \text{ primitive}$$

$$M \vDash_{PM} (\neg R)^+ \text{ iff } M \vDash_{PM} R^-$$

$$M \vDash_{PM} (\neg R)^- \text{ iff } M \vDash_{PM} R^+$$

$$M \vDash_{PM} (R \vee Q)^+ \text{ iff } M \vDash_{PM} R^+ \text{ or } M \vDash_{PM} Q^+$$

$$M \vDash_{PM} (R \vee Q)^- \text{ iff } M \vDash_{PM} R^- \text{ and } M \vDash_{PM} Q^-$$

Conjunction, implication and double implication are defined from negation and disjunction in the usual way. It is easy to build up partial models A or L in which the law of excluded middle fails, that is, we have neither $M \vDash_{PM} S^+$ nor $M \vDash_{PM} S^-$ even at the level of atomic propositional symbols.

In case L is a first-order language (containing only relational symbols), a partial model for it is defined analogously. That is, each relational symbol R is assigned both an extension R^{M+} and a counterextension R^{M-}. The inductive definition of satisfaction (with respect to an assignment) is then defined analogously with the propositional case. The new clauses are

$(M,g)\vDash_{PM}(\exists x\phi)^+$ iff there is an $a\in dom(M)$ such that $(M,g\cup\{(x,a)\})\vDash_{PM}\phi^+$

$(M,g)\vDash_{PM}(\exists x\phi)^-$ iff for all $a\in dom(M)$ $(M,g\cup\{(x,a)\})\vDash_{PM}\phi^-$

Let L be a first-order language in an arbitrary signature, and N a partial model for it. A game-theoretical interpretation for L is analogous to the game-theoretical interpretation on classical models for L. The only difference is with atomic formulas. That is, clause (i) of Definition 1 in section 1 is now replaced by

(1) $G(R(t_1,\ldots,t_n),M,g)$ contains no moves. If $(t_1^g,\ldots,t_n^g)\in R^{M+}$, then the player who is the verifier wins $G(R(t_1,\ldots,t_n),M,g)$; If $(t_1^g,\ldots,t_n^g)\in R^{M-}$, then the player who is the falsifier wins $G(R(t_1,\ldots,t_n),M,g)$.

The clause for identity is the same as in the case of classical models. The proof of the next proposition is straightforward.

Proposition 1. For any first-order formula of L, partial model M of L and assignment g in M:

$$(M,g)\vDash_{PM}\phi^+ \text{ iff } (M,g)\vDash_{GTS}\phi^+ \qquad (*)$$

$$(M,g)\vDash_{PM}\phi^- \text{ iff } (M,g)\vDash_{GTS}\phi^-. \qquad (**)$$

Partiality in the models can be combined with partiality of information, that is, we can give a game-theoretical interpretation for an IF first-order language L and a partial model M of L. All the relevant definitions remain the same.

4. Kripke's result

Let L be a first-order language in the signature of Peano Arithmetic. We enrich L with a truth-predicate Tr and denote the resulting language by L^+. We expand the classsical model N of L to a classical model (N,E) of L^+ (E is the interpretation of Tr). We get a partial model $(N,(E,D))$ of L^+ by adding a set D disjoint from E to be the counterextension of Tr. Intuitively, E is intended to consist of the names (Gödel numbers) of the sentences of L^+ which are true in $(N, (E, D))$, and D of those sentences (and possibly other individuals in N) to which Tr does not apply. Notice that in this partial model, the counterextensions of all the predicate symbols, except Tr, are the complements of the respective extensions.

Tarski's classical result is that we cannot find a classical model (N, E) of L^+ such that E is the set of the Gödel numbers of the sentences of L^+ true in (N, E).

Kripke's well-known result is that we can find a partial model which has the analogous property, that is, we can find (E, D) such that

$$(N, (E, D)) \vDash_{PM} (Tr({}^{\ulcorner}\phi{}^{\urcorner}))^+ \text{ iff } (N, (E, D)) \vDash_{PM} \phi^+ \tag{A.4.1}$$

$$(N, (E, D)) \vDash_{PM} (Tr({}^{\ulcorner}\phi{}^{\urcorner}))^- \text{ iff } (N, (E, D)) \vDash_{PM} \phi^- \tag{A.4.2}$$

Such an (E, D) is called a *fixed point* for L^+. The existence of a minimal fixed point can be proved by a straightforward transfinite construction: We let

$$E_\alpha = \{ {}^{\ulcorner}\phi{}^{\urcorner} \colon (N, \bigcup_{\beta < \alpha} E_\beta, \bigcup_{\beta < \alpha} D_\beta) \vDash_{PM} \phi^+ \}$$

$$D_\alpha = \{ n \in N \colon \neg st(n) \} \cup \{ {}^{\ulcorner}\phi{}^{\urcorner} \colon (N, \bigcup_{\beta < \alpha} E_\beta, \bigcup_{\beta < \alpha} D_\beta) \vDash_{PM} \phi^- \}$$

We then set $E_\infty = \bigcup_{\alpha \in Or} E_\alpha$ and $D_\infty = \bigcup_{\alpha \in Or} D_\alpha$. By cardinality considerations, there are E_λ and D_λ such that for all $\kappa > \lambda$, $E_\lambda = E_\kappa$ and $D_\lambda = D_\kappa$. The pair (E_λ, D_λ) is the minimal fixed point.

The fixed point (E, D) for Kripke's first-order language L^+ has been axiomatically characterized by Feferman (1984):

Let KF(Tr) (Kripke–Feferman axioms) be the conjunction of the following sentences in the language L^+:

$$\forall x (Tr(x) \rightarrow st(x)) \tag{A.4.3}$$

(ϕ is an atomic sentence of L or the negation of an atomic sentence of $L \rightarrow (Tr({}^{\ulcorner}\phi{}^{\urcorner}) \leftrightarrow \phi)$) $\tag{A.4.4}$

$$\forall x (ct(x) \rightarrow (Tr({}^{\ulcorner}Tr(x){}^{\urcorner}) \leftrightarrow Tr(x))) \tag{A.4.5}$$

$$\forall x (ct(x) \rightarrow (Tr({}^{\ulcorner}\neg Tr(x){}^{\urcorner}) \leftrightarrow Tr({}^{\ulcorner}\neg x{}^{\urcorner}))) \tag{A.4.6}$$

$$\forall x \forall y ((st(x) \wedge st(y)) \rightarrow (Tr({}^{\ulcorner}(x \vee y){}^{\urcorner}) \leftrightarrow (Tr(x) \vee Tr(y)))) \tag{A.4.7}$$

$$\forall x \forall y ((st(x) \wedge st(y)) \rightarrow (Tr({}^{\ulcorner}\neg (x \vee y){}^{\urcorner}) \leftrightarrow (Tr({}^{\ulcorner}\neg x{}^{\urcorner}) \wedge Tr({}^{\ulcorner}\neg y{}^{\urcorner})))) \tag{A.4.8}$$

$$\forall x(\mathrm{st}(x) \to (\mathrm{Tr}(^{\ulcorner}\neg\,^{\ulcorner}\neg\,x^{\urcorner\urcorner}) \leftrightarrow \mathrm{Tr}(x))) \tag{A.4.9}$$

$$\forall y(\mathrm{fl}(y) \wedge \mathrm{st}(^{\ulcorner}\exists xy^{\urcorner}) \to (\mathrm{Tr}(^{\ulcorner}\exists xy^{\urcorner}) \leftrightarrow \exists x\mathrm{Tr}(\mathrm{Sub}(y,{}^{\ulcorner}x^{\urcorner},x))) \tag{A.4.10}$$

$$\forall y(\mathrm{fl}(y) \wedge \mathrm{st}(^{\ulcorner}\neg\,\exists xy^{\urcorner}) \to (\mathrm{Tr}(^{\ulcorner}\neg\,\exists xy^{\urcorner}) \leftrightarrow \forall x\mathrm{Tr}(\mathrm{Sub}(^{\ulcorner}\neg\,y^{\urcorner},{}^{\ulcorner}x^{\urcorner},x)))$$
$$\tag{A.4.11}$$

$$\neg\,\exists x(\mathrm{Tr}(x) \wedge \mathrm{Tr}(^{\ulcorner}\neg\,x^{\urcorner})) \tag{A.4.12}$$

Feferman (1984) proved the following:

Theorem 1. Let E be a subset of N. Then the classical model (N, E) is a model of KF(Tr) iff $(N, E, \{\text{nonsentences of } L^+\} \cup \{^{\ulcorner}\phi^{\urcorner}: {}^{\ulcorner}\neg\,\phi^{\urcorner}\in E\})$ is a fixed point for L^+. ∎

5. GTS and partial models

It is time to relate our result on the definability of truth to Kripke's result. But first, it might be interesting to point out the similarities and the dissimilarities between the two kinds of partiality.

In the case of partial predicates (Kripke), we start with a *new primitive predicate* Tr. Then we can find a partial interpretation (E, D) for it such that "truth-in-$(N, (E, D))$" (in the sense of partial models) is defined by Tr. In the case of partiality of information (GTS), we are able to find a *formula* $\theta(x)$ which defines "truth-in-N" (in the sense of GTS).

The truth-defining formula $\theta(x)$ cannot be replaced by a predicate symbol by the standard technique, that is, by means of an explicit definition

$$N \vDash_{\mathrm{GTS}} \forall x(S(x) \leftrightarrow \theta(x))^+, \tag{A.5.1}$$

because, given the fact that $\theta(x)$ does not satisfy the law of excluded middle, the predicate S would have to be partially interpreted. If we want to reduce $\theta(x)$ to a predicate, we then have to combine the two kinds of partiality, that is, to have an IF language which is game-theoretically interpreted, and in addition to allow for the predicate symbols of L to be partially interpreted. We indicated at the end of section 3 how this can be done.

So let L be an IF language (in the signature of Peano Arithmetic) and let us add to it a predicate symbol Tr. We denote the resulting language by L^+. Then, by a transfinite construction similar to that for partial models in the preceding section, it can be shown that there is a fixed point (E, D) for L^+:

$$(N, (E, D)) \vDash_{\text{GTS}} (\text{Tr}(\ulcorner\phi\urcorner))^+ \text{ iff } (N, (E, D)) \vDash_{\text{GTS}} \phi^+ \tag{A.5.2}$$

$$(N, (E, D)) \vDash_{\text{GTS}} (\text{Tr}(\ulcorner\phi\urcorner))^- \text{ iff } (N, (E, D)) \vDash_{\text{GTS}} \phi^- \tag{A.5.3}$$

A syntactical characterization of a fix point, in the style of Feferman can be obtained by adding to the Kripke–Feferman axioms from the preceding section the sentences

$$\forall y((\text{fl}(y) \wedge \text{st}(\ulcorner G_{1,1}^{1,1} x_1 y_1 z_1 t_1 y\urcorner)) \to (\text{Tr}(\ulcorner G_{1,1}^{1,1} x_1 y_1 z_1 t_1 y\urcorner) \leftrightarrow$$

$$\exists f g \forall x_1 z_1 \, \text{Tr}(\text{Sub}(\text{Sub}(\text{Sub}(\text{Sub}(y, \ulcorner y_1\urcorner, f(x_1)), \ulcorner t_1\urcorner, g(z_1)),$$

$$\ulcorner x_1\urcorner, x_1), \ulcorner z_1\urcorner, z_1)))) \tag{A.5.4}$$

$$\forall y((\text{fl}(y) \wedge \text{st}(\ulcorner \neg G_{1,1}^{1,1} x_1 y_1 z_1 t_1 y\urcorner)) \to (\text{Tr}(\ulcorner \neg G_{1,1}^{1,1} x_1 y_1 z_1 t_1 y\urcorner) \leftrightarrow$$

$$\exists f x \forall y_1 w_1 \text{Tr}(\text{Sub}(\text{Sub}(\text{Sub}(\text{Sub}(\ulcorner \neg y\urcorner, \ulcorner x_1\urcorner, x_1), \ulcorner z_1\urcorner, f(y_1)),$$

$$\ulcorner y_1\urcorner, y_1), \ulcorner t_1\urcorner, t_1)))). \tag{A.5.5}$$

∎

We denote the result by KF*(Tr). We then have

Theorem 1. Let L be an IF language (in the signature of Peano Arithmetic), N the standard model of L, and Tr a new predicate symbol. Then for any $A \subseteq \text{dom}(N)$ we have

(*) $(N, A) \vDash \text{KF}^*(\text{Tr})$ iff $(N, A, \{\text{nonsentences}\} \cup \{\ulcorner\phi\urcorner : \ulcorner \neg \phi\urcorner \in A\})$ is a fixed point for $L \cup \{\text{Tr}\}$.

Proof: ⇒
Let $K = \{\text{nonsentences of } L \cup \{\text{Tr}\}\} \cup \{\ulcorner\phi\urcorner : \ulcorner \neg \phi\urcorner \in A\}$. We assume $(N, A) \vDash \text{KF}^*(\text{Tr})$, and then show by induction on the complexity of ϕ that (A, K) is a fixed point, that is

$$(N, (A, K)) \vDash_{\text{GTS}} \phi^+ \text{ iff } \ulcorner\phi\urcorner \in A \tag{A.5.6}$$

$$(N, (A, K)) \vDash_{\text{GTS}} \phi^- \text{ iff } \ulcorner\phi\urcorner \in K \tag{A.5.7}$$

For all the first-order clauses, the proof is identical with that given for Feferman for partial models in the first-order case. So it is enough to consider the two additional clauses related to IF languages. However, in order to make the connection with the proof given by

Feferman more transparent, we shall also review the clause for negation.

Suppose ϕ is $\neg\,\psi$, and $(N, (A, K)) \vDash_{GTS} (\neg\,\psi)^+$. Then, by the definition of a fixed point, $^\ulcorner\neg\,\psi^\urcorner \in A$. Conversely, assume $^\ulcorner\neg\,\psi^\urcorner \in A$. Then by the definition of K, $^\ulcorner\psi^\urcorner \in K$, and by the inductive hypothesis, $(N, (A, K)) \vDash_{GTS} \psi^-$. Hence, by Definition 2 (i) of section 1, $(N, (A, K)) \vDash_{GTS} (\neg\,\psi)^+$.

In order to prove (7), assume $(N, (A, K)) \vDash_{GTS} (\neg\,\psi)^-$. Hence, by Definition 2 (ii), $(N, (A, K)) \vDash_{GTS} \psi^+$. By the inductive hypothesis, $^\ulcorner\psi^\urcorner \in A$, hence $(N, A) \vDash \text{Tr}(^\ulcorner\psi^\urcorner)$. By (8) of KF(Tr), $(N, A) \vDash (\text{Tr}(^\ulcorner{}^\ulcorner\neg\,\psi^\urcorner{}^\urcorner))$. Hence $^\ulcorner\neg\,{}^\ulcorner\neg\,\psi^\urcorner{}^\urcorner \in A$, and thus $^\ulcorner\neg\,\psi^\urcorner \in K$. The converse is similar.

Suppose that ϕ is $G^{1;1}_{1;1} x_1 y_1 z_1 t_1 \psi$ and assume $(N, (A, K)) \vDash_{GTS} G^{1;1}_{1;1} x_1 y_1 z_1 t_1 \psi^+$. Then there are functions f and g such that for all natural numbers n and m: $(N, (A, K)) \vDash_{GTS} \psi(n, f(n), m, g(m))^+$. By the inductive hypothesis, $^\ulcorner\psi(n, f(n), m, g(m))^\urcorner \in A$. We thus showed that there are functions f and g such that for all n and m $\text{Sub}(\text{Sub}(\text{Sub}(\text{Sub}(^\ulcorner\psi^\urcorner, {}^\ulcorner y_1^\urcorner, f(n)), {}^\ulcorner t_1^\urcorner, g(m)), {}^\ulcorner x_1^\urcorner, n), {}^\ulcorner z_1^\urcorner, m) \in A$. But $(N, A) \vDash \text{KF}^*(\text{Tr})$, hence by (A.5.4) we get $(N, A) \vDash \text{Tr}(^\ulcorner G^{1;1}_{1;1} x_1 y_1 z_1 t_1 \psi^\urcorner)$, that is, $^\ulcorner G^{1;1}_{1;1} x_1 y_1 z_1 t_1 \psi^\urcorner \in A$.

Conversely, assume $^\ulcorner G^{1;1}_{1;1} x_1 y_1 z_1 t_1 \psi^\urcorner \in A$. Then (since A is the interpretation of Tr) $(N, A) \vDash \text{Tr}(^\ulcorner G^{1;1}_{1;1} x_1 y_1 z_1 t_1 \psi^\urcorner)$. But (N, A) satisfies KF*(Tr), therefore there are functions f and g such that $(N, A) \vDash \text{Tr}(^\ulcorner\psi(m, p, n, s)^\urcorner)$, for all quadruples (m, p, n, s) such that $p = f(m)$, and $s = g(n)$. Hence $^\ulcorner\psi(m, p, n, s)^\urcorner \in A$. By the inductive hypothesis $(N, (A, K)) \vDash_{GTS} \psi(m, p, n, s)^+$, for all quadruples (m, p, n, s) such that $p = f(m)$, and $s = g(n)$. But then it is easy to see that Myself has a winning strategy in $G(G^{1;1}_{1;1} x_1 y_1 z_1 t_1 \psi, (N, (A, K))$. Therefore $(N, (A, K)) \vDash_{GTS} G^{1;1}_{1;1} x_1 y_1 z_1 t_1 \psi^+$. This proves (A.5.6).

In order to prove (A.5.7), assume $(N, (A, K)) \vDash_{GTS} G^{1;1}_{1;1} x_1 y_1 z_1 t_1 \psi^-$. Then, by the definition of the fixed point, $^\ulcorner G^{1;1}_{1;1} x_1 y_1 z_1 t_1 \psi^\urcorner \in K$.

Conversely, assume $^\ulcorner G^{1;1}_{1;1} x_1 y_1 z_1 t_1 \psi^\urcorner \in K$. Then, by the definition of K, $^\ulcorner G^{1;1}_{1;1} x_1 y_1 z_1 t_1 \psi^\urcorner \in A$. Hence $(N, A) \vDash \text{Tr}(^\ulcorner\neg\, G^{1;1}_{1;1} x_1 y_1 z_1 t_1 \psi^\urcorner)$. By (A.5.5), there is an m and a function f such that for all p and s we have $(N, A) \vDash \text{Tr}(^\ulcorner\neg\,\psi(m, p, f(p), s)^\urcorner)$. Hence $^\ulcorner\neg\,\psi(m, p, f(p), s)^\urcorner \in A$. By the inductive hypothesis, (*) $(N, (A, K)) \vDash_{GTS} \psi(m, p, n, s)^-$. But then it is easy to see that Nature has a winning strategy in $G(G^{1;1}_{1;1} x_1 y_1 z_1 t_1 \psi, (N, (A, K)))$. In his first move, let Nature choose an m as above. Then for any arbitrary choice p made by Myself, let Nature choose $f(p)$. We then know by (*) above, that for whatever s chosen by Myself, Nature has a winning strategy in $G(\psi, (N, (A, K), m, p, n, s))$. ◼

It is straightforward to show that, if (A, K) is a fixed point, then the axioms KF*(Tr) are true in (N, A).

The result we reached here is stronger than Kripke's given the fact that the partially interpreted IF languages are much stronger than partially interpreted first-order languages. ∎

* I am grateful to Jouko Väänänen, Taneli Huuskonen, and Kerkko Luosto for useful suggestions concerning the material of this appendix.

References

Baldus, Richard, 1928. "Zur Axiomatik der Geometrie 1: Über Hilberts Vollständigkeitsaxiom," *Mathematische Annalen* 100, 321–333.

Barwise, Jon, 1979. "On Branching Quantifiers in English," *Journal of Philosophical Logic* 8, 47–80.

Barwise, Jon, 1985. "Model-Theoretic Logics: Background and Aims," in Barwise and Feferman 1985, 3–23.

Barwise, Jon, and Solomon Feferman (eds.), 1985. *Model-Theoretic Logics*, Springer, New York.

Benacerraf, Paul, 1973. "Mathematical Truth," *Journal of Philosophy* 70, 661–680; reprinted in Benacerraf and Putnam (1983), 403–420.

Benacerraf, Paul, and Hilary Putnam (eds.), 1964. *Philosophy of Mathematics*, Prentice-Hall, Englewood Cliffs, NJ. Second ed. 1983, Cambridge University Press, Cambridge.

van Benthem, Johan, and Alice ter Meulen (eds.), 1996. *Handbook of Logic and Language*, Elsevier, Amsterdam.

Blass, Andreas, and Yuri Gurevich, 1986. "Henkin Quantifiers and Complete Problems," *Annals of Pure and Applied Logic* 32, 1–16.

Blumenthal, Otto, 1935. "Lebensgeschichte [Hilberts]," in Hilbert (1935) 388–429.

Boltzmann, Ludwig, 1905. *Populäre Schriften*, Verlag von Johann Ambrosius Baerth, Leipzig.

Boolos, George, 1971. "The Iterative Conception of Set," *Journal of Philosophy* 68, 215–232. Reprinted in Benacerraf and Putnam (1983), 486–502.

Bottazzini, Umberto, 1986. *The "Higher Calculus": A History of Real and Complex Analysis from Euler to Weierstrass*, Springer, Berlin.

Brouwer, L. E. J., 1975. *Collected Works* I, ed. by A. Heyting, North-Holland, Amsterdam.

Carnap, Rudolf, 1934. *Logische Syntax der Sprache*, Springer, Wien.

Chomsky, Noam, 1977. *Essays on Form and Interpretation*, North-Holland, Amsterdam.

Chomsky, Noam, 1981. *Lectures on Government and Binding*, Foris, Dordrecht.

271

272   REFERENCES

Chomsky, Noam, 1986. *Knowledge of Language*, Praeger, New York.

Cohen, Paul, 1966. *Set Theory and the Continuum Hypothesis*, Benjamin, New York.

Cummins, Robert, 1989. *Meaning and Mental Representation*, MIT Press, Cambridge, MA.

Dauben, J. W., 1979. *Georg Cantor: His Mathematics and Philosophy of the Infinite*, Harvard University Press, Cambridge, MA.

Davidson, Donald, 1965. "Theories of Meaning and Learnable Languages," in Yehoshua Bar-Hillel (ed.), *Logic, Methodology and Philosophy of Science, Proceedings of the 1964 International Congress*, North-Holland, Amsterdam, 383–394.

Dawar, Anuj, and Lauri Hella, forthcoming. "The Expressive Power of Finitely Many Generalized Quantifiers."

Devlin, Keith, 1984. *Constructibility*, Springer, New York.

Dummett, Michael, 1977. *Elements of Intuitionism*, Clarendon Press, Oxford.

Dummett, Michael, 1978. *Truth and Other Enigmas*, Duckworth, London.

Dummett, Michael, 1991. *The Logical Basis of Metaphysics*, Harvard University Press, Cambridge, MA.

Dummett, Michael, 1993. *The Seas of Language*, Clarendon Press, Oxford.

Easton, W. B., 1964. Powers of Regular Cardinals, Ph. D. Dissertation, Princeton University.

Ebbinghaus, H. D., J. Flum and W. Thomas, 1984. *Mathematical Logic*, Springer, New York.

Ekeland, Ivar, 1988. *Mathematics and the Unexpected*, University of Chicago Press, Chicago.

Enderton, H. B., 1970. "Finite Partially Ordered Quantifiers," *Zeitschrift für Mathematische Logik und Grundlagen der Mathematik* 16, 393–397.

Engdahl, Elisabeth, 1986. *Constituent Questions*, D. Reidel, Dordrecht.

Erdös, P., A. Hajnal, A. Máté, and R. Rado, 1984. *Combinatorial Set Theory: Partial Relations for Cardinals*, North-Holland, Amsterdam.

Euler, Leonhard, 1988. *Introduction to Analysis of the Infinite I*, translated by John D. Blanton, Springer, Berlin. (Original 1748.)

Feferman, Solomon, 1984. "Towards useful Type-Free Theories, I," *Journal of Symbolic Logic* 49, 75–111.

Feferman, Solomon, 1993. "What Rests on What? The Proof-Theoretic Analysis of Mathematics," in J. Czermak (ed.), *Philosophy of Mathematics: Proceedings of the 15th International Wittgenstein Symposium*. Hölder–Pichler–Tempsky, Vienna, pp. 147–171.

Felgner, Ulrich, 1971. "Comparison of the Axioms of Logical and Universal Choice," *Fundamenta Mathematicae* 71, 43–62.

Fenstad, Jens–Erik, 1971. "The Axiom of Determinateness," in J.-E. Fenstad (ed.), *Proceedings of the Second Scandinavian Logic Symposium*, North-Holland, Amsterdam.

Fenstad, Jens-Erik, 1996. "Partiality," in van Benthem and ter Meulen, 1996.

Fernández Moreno, Luis, 1992. *Wahrheit und Korrespondenz bei Tarski*, Königshausen & Neumann, Würzburg.

Fine, Terrence L., 1973. *Theories of Probability*, Academic Press, New York.

Frege, Gottlob, 1879. *Begriffsschrift, eine der aritmetischen nachgebildete Formelsprache des reinen Denkens*, Nebert, Halle.

Frege, Gottlob, 1980. *Philosophical and Mathematical Correspondence*, Basil Blackwell, Oxford.

Freudenthal, Hans, 1957. "Zur Geschichte der Grundlagen der Geometrie," *Nieuw Archief voor Wiskunde* 5, 105–142.

Gaifman, Haim, 1975. "Global and Local Choice Functions," *Israel Journal of Mathematics* 22, 257–265.

Gödel, Kurt, 1930. "Die Vollständigkeit der Axiome des logischen Funktionenkalküls," *Monatshefte für Mathematik und Physik* 37, 340–360; reprinted in English translation in Gödel 1986, 102–123 (see also 44–101).

Gödel, Kurt, 1931. "Über formal unentscheidbare Sätze der *Principia Mathematica* und verwandter Systeme 1," *Monatshefte für Mathematik und Physik* 38, 173–198; reprinted in English translation in Gödel 1986, 144–195.

Gödel, Kurt, 1940. *The Consistency of the Axiom of Choice and of the Generalized Continuum Hypothesis (Annals of Mathematical Studies 3)*, Princeton University Press, Princeton.

Gödel, Kurt, 1947. "What Is Cantor's Continuum Problem?" *American Mathematical Monthly* 54, 515–525. Reprinted in an expanded form in Benacerraf and Putnam (1983), 470–485.

Gödel, Kurt, 1958. "Über eine bisher noch nicht benützte Erweiterung des finiten Standpunktes," *Dialectica* 12, 280–287; reprinted in English translation in Gödel 1990, 241–251.

Gödel, Kurt, 1986. *Collected Works*, vol. 1, Oxford University Press, New York.

Gödel, Kurt, 1990. *Collected Works*, vol. 2, Oxford University Press, New York.

Goldfarb, Warren, 1979. "Logic in the Twenties: The Nature of the Quantifier," *The Journal of Symbolic Logic* 44, 351–368.

Goldfarb, Warren, 1989. "Russell's Reasons for Ramification," in C. Wade Savage and C. Anthony Anderson (eds.), *Rereading Russell (Minnesota Studies in Philosophy of Science 12)*, University of Minnesota Press, Minneapolis, 24–40.

Grabiner, Judith V., 1981. *The Origins of Cauchy's Rigorous Calculus*, MIT Press, Cambridge.

Graham, Ronald L., Bruce L. Rotschild, and Joel H. Spencer, 1990. *Ramsey Theory*, 2nd ed., John Wiley & Sons, New York.

Graham, Ronald L., and Joel H. Spencer, 1990. "Ramsey Theory," *Scientific American* 262, no. 1 (July 1990), 112–117.

Hallett, Michael, 1984. *Cantorian Set Theory and Limitation of Size*, Clarendon Press, Oxford.

Heim, Irene R., 1982. *The Semantics of Definite and Indefinite Noun Phrases*, Dissertation at the University of Massachusetts, Amherst, MA.

Henkin, Leon, 1950. "Completeness in the Theory of Types," *Journal of Symbolic Logic* 15, 81–91.

Henkin, Leon, 1961. "Some Remarks on Infinitely Long Formulas," in (no editor given) *Infinitistic Methods*, Pergamon Press, Oxford, 167–183.

Heyting, Arend, 1931. "Die intuitionistische Grundlegung der Mathematik," *Erkenntnis* 2, 106–115; English translation in Benacerraf and Putnam, 1983, 52–61.

Heyting, Arend, 1956. *Intuitionism: An Introduction*, North-Holland, Amsterdam.

Hilbert, David, 1899. "Grundlagen der Geometrie," in B. G. Teubner (ed.), *Festschrift zur Freier der Enthüllung des Gauss-Weber Denkmals in Göttingen*, Leipzig, pp. 3–92.

Hilbert, David, 1900. "Über den Zahlbegriff," *Jahresbericht des Deutschen Mathematiker Vereinigung* 8, 180–184.

Hilbert, David, 1918. "Axiomatisches Denken," *Mathematische Annalen* 78, 405–415.

Hilbert, David, 1922. "Neubegründung der Mathematik. Erste Mitteilung," *Abhandlungen aus dem Mathematischen Seminar der Hamburg Universität* 1, 157–177.

Hilbert, David, 1925. "Über das Unendliche," *Mathematische Annalen* 95, 161–190. English translation in Benacerraf and Putnam 1983, 183–201.

Hilbert, David, 1935. *Gesammelte Abhandlungen* 3, Springer, Berlin.

Hilbert, David, and W. Ackermann, 1928. *Grundzüge der theoretischen Logik*, Springer, Berlin.

Hilpinen, Risto, 1983. "On C. S. Peirce's Theory of the Proposition: Peirce as a Precursor of Game-Theoretical Semantics," in Eugene Freeman, (ed.), *The Relevance of Charles Peirce*, The Hegeler Institute, La Salle, Illinois, 264–270.

Hintikka, Jaakko, 1973. *Logic, Language-Games and Information: Kantian Themes in the Philosophy of Logic*, Clarendon Press, Oxford.

Hintikka, Jaakko, 1974. "Quantifiers vs. Quantification Theory," *Linguistic Inquiry* 5, 153–177.

Hintikka, Jaakko, 1975. "Impossible Possible Worlds Vindicated," *Journal of Philosophical Logic* 4, 475–484.

Hintikka, Jaakko, 1976a. "Quantifiers in Logic and Quantifiers in Natural Languages," in Stephan Körner (ed.), *Philosophy of Logic*, Basil Blackwell, Oxford.

Hintikka, Jaakko, 1976b. "The Prospects of Convention T," *Dialectica* 30, 61–63.

Hintikka, Jaakko, 1979. "Quantifiers in Natural Languages." In Saarinen (1979), 81–117.

Hintikka, Jaakko, 1985, "A Spectrum of Logics for Questioning," *Philosophica* 35, 135–150.

Hintikka, Jaakko, 1987. "Game-Theoretical Semantics as a Synthesis of Truth-Conditional and Verificationist Theories of Meaning," in Ernest LePore (ed.), *New Directions in Semantics*, Academic Press, London.

Hintikka, Jaakko, 1988a. "What Is the Logic of Experimental Inquiry?" *Synthese* 74, 173–190.

Hintikka, Jaakko, 1988b. "On the Development of the Model-Theoretical Viewpoint in Logical Theory," *Synthese* 77, 1–36.

Hintikka, Jaakko, 1988c. "Model Minimization – An Alternative to Circumscription," *Journal of Automated Reasoning* 4, 1–13.

Hintikka, Jaakko, 1989. "Logical Form and Linguistic Theory," in Alex George (ed.), *Reflections on Chomsky*, Basil Blackwell, Oxford, 41–57.

Hintikka, Jaakko, 1990. "Paradigms for Language Theory," *Acta Philosophica Fennica* 49, 181–209.

Hintikka, Jaakko, 1992. "Different Constructions in Terms of 'Knows'", in Jonathan Dancy and Ernest Sosa, eds., *A Companion to Epistemology*, Basil Blackwell, Oxford, 99–104.

Hintikka, Jaakko, 1993a. "New Foundations for Mathematical Theories," in J.Oikkonen and J. Väänänen, eds., *Logic Colloquium '90 (Lecture Notes in Logic* 2), Springer, Berlin, 122–144.

Hintikka, Jaakko, 1993b. "The Original *Sinn* of Wittgenstein's Philosophy of Mathematics," in Klaus Puhl (ed.), *Wittgenstein's Philosophy of Mathematics*, Hölder–Pichler–Tempsky, Vienna, 24–51.

Hintikka, Jaakko, 1993c. "Gödel's Functional Interpretation in a Wider Perspective," in H.-D. Schwabl (ed.), *Yearbook 1991 of the Kurt Gödel Society*, Vienna, 5–39.

Hintikka, Jaakko, 1995a. "The Standard vs. Nonstandard Distinction: A Watershed in the Foundations of Mathematics." in Jaakko Hintikka (ed.), *From Dedekind to Gödel*, Kluwer Academic, Dordrecht, 21–44.

Hintikka, Jaakko, 1995b. "What Is Elementary Logic? Independence-Friendly Logic as the True Core Arc of Logic," in Kostas Gavroglu et al. (eds.), *Physics, Philosophy, and the Scientific Community*, Kluwer Academic, Dordrecht, 301–326.

Hintikka, Jaakko, 1996. "Hilbert Vindicated?" *Synthese* (forthcoming).

Hintikka, Jaakko, and Jack Kulas, 1983. *The Game of Language*, D. Reidel, Dordrecht.

Hintikka, Jaakko, and Jack Kulas, 1985. *Anaphora and Definite Descriptions: Two Applications of Game-Theoretical Semantics*, D. Reidel, Dordrecht.

Hintikka, Jaakko, and Veikko Rántala, 1976. "A New Approach to Infinitary Languages," *Annals of Mathematical Logic*, 10, 95–115.

Hintikka, Jaakko, and Unto Remes, 1974. *The Method of Analysis*, D. Reidel, Dordrecht.

Hintikka, Jaakko, and Gabriel Sandu, 1991. *On the Methodology of Linguistics: A Case Study*, Basil Blackwell, Oxford.

Hintikka, Jaakko, and Gabriel Sandu, 1994. "What Is a Quantifier?," *Synthese* 98, 113–129.

Hintikka, Jaakko, and Gabriel Sandu, 1995. "What Is the Logic of Parallel Processing?" *International Journal of Foundations of Computer Science*, 6, 27–49.

Hintikka, Jaakko, and Gabriel Sandu, 1996. "Game-theoretical Semantics," in van Benthem and ter Meulen, 1996.

Hintikka, Merrill B., and Jaakko Hintikka, 1986, *Investigating Wittgenstein*, Basil Blackwell, Oxford.

Horn, Laurence R., 1989. *A Natural History of Negation*, University of Chicago Press.

Hughes, R. I. G., 1989. *The Structure and Interpretation of Quantum Mechanics*, Harvard University Press, Cambridge, MA.

Jackendoff, Ray S., 1972. *Semantic Interpretation in Generative Grammar*, MIT Press, Cambridge, MA.

Jeroslow, R. G., 1973. "Redundancies in the Hilbert-Bernays Derivability Conditions for Gödel's Second Incompleteness Theorem," *The Journal of Symbolic Logic* 38, 359–367.

Jones, James P., 1974. "Recursive Undecidability – An Exposition," *The American Mathematical Monthly* 81, 724–738.

Kanamori, Akihiro, 1994. *The Higher Infinite*, Springer-Verlag, Berlin.

Kanamori, Akihiro, and M. Magidor, 1978. "The Evolution of Large Cardinal Axioms in Set Theory," in G. H. Müller and Dana Scott (eds.), *Higher Set Theory* (*Lecture Notes in Mathematics* 699), Springer, New York, 99–275.

Keenan, Edward, and Dag Westerstahl, 1996. "Generalized Quantifiers," in van Benthem and ter Meulen, 1996.

Kleene, Stephen C., 1952. *Introduction to Metamathematics*, Van Nostrand, New York.

Kleinberg, Eugene M., 1977. *Infinitary Combinatorics and the Axiom of Determinateness* (*Lecture Notes in Mathematics* 612), Springer, Berlin.

Kreisel, Georg, 1953. "A Note on Arithmetic Models for Consistent Formulae of the Predicate Calculus II," in the *Proceedings of the XI International Congress of Philosophy* 14, Amsterdam & Louvain, 37–47.

Kripke, Saul, 1975. "Outline of a Theory of Truth," *Journal of Philosophy* 72, 690–716; reprinted in Martin 1984, 53–81.

Krynicki, Michael, 1993. "Hierarchies of Partially Ordered Connectives and Quantifiers," *Mathematical Logic Quarterly* 39, 287–294.

Krynicki, Michael, and Jouko Väänänen, 1989. "Henkin and Function Quantifiers," *Annals of Pure and Applied Logic* 43, 273–292.

Kunen, Kenneth, 1980. *Set Theory: An Introduction to Independence Proofs*, North-Holland, Amsterdam.

Kusch, Martin, 1989. *Language as the Universal Medium vs. Logic as Calculus: A Study of Husserl, Heidegger and Gadamer*, Kluwer, Dordrecht.

Lindström, Per, 1969. "On Extensions of Elementary Logic," *Theoria* 35, 1–11.

Lovejoy, A. O., 1936. *The Great Chain of Being*, Harvard University Press, Cambridge, MA.

Mancosu, Paolo, 1991. "On the Status of Proofs by Contradiction in the Seventeenth Century," *Synthese* 88, 15–41.

Martin, Robert L. (ed.), 1978. *The Paradox of the Liar*, Ridgeview, Atascadero, CA.

Martin, Robert L. (ed.), 1984. *Recent Essays on Truth and the Liar Paradox*, Clarendon Press, Oxford.

Martin–Löf, Per, 1966. "The Definition of Random Sequences," *Information and Control* 6, 602–619.

Martin–Löf, Per, 1970. "On the Notion of Randomnesss," in A. Kino, J. Myhill, and R. E. Vesley (eds.), *Intuitionism and Proof Theory*, North-Holland, Amsterdam, 73–78.

Martin–Löf, Per, 1984. *Intuitionistic Type Theory*, Bibliopolis, Napoli.

Matiyasevich, Yuri V., 1993. *Hilbert's Tenth Problem*, MIT Press, Cambridge, MA.

McGee, Vann, 1991. *Truth, Vagueness and Paradox*, Hackett, Indianapolis.

Mendelson, Elliott, 1987. *Introduction to Mathematical Logic*, 3rd ed., Wadsworth & Brooks/Cole, Monterey, CA.

Moore, Gregory H., 1982. *Zermelo's Axiom of Choice*, Springer, Berlin.

Moore, Gregory H., 1988. "The Emergence of First-Order Logic," in William Aspray and Philip Kitcher (eds.), *History and Philosophy of Modern Mathematics* (*Minnesota Studies in the Philosophy of Science* XI), University of Minnesota Press, Minneapolis, 95–135.

Moore, Gregory H., 1994. "Logic and Set Theory," in I. Grattan-Guinness (ed.), *Companion Encyclopedia of the History and Philosophy of the Mathematical Sciences*, vol. 1, Routledge, London, 635–643.

Moreno, *see* Fernández Moreno.

Morris, Charles, 1938. *Foundations of the Theory of Signs* (International Encyclopedia of Unified Science 1, no. 2), University of Chicago Press, Chicago.

Moschovakis, Y. N., 1974. *Elementary Induction on Abstract Structures*, North-Holland, Amsterdam.

Mostowski, Andrzej, 1955. "A Formula with no Recursively Enumerable Model," *Fundamenta Mathematicae* 42, 125–140.

Mostowski, Andrzej, 1965. *Thirty Years of Foundational Studies* (Acta Philosophica Fennica 17), Helsinki.

von Neumann, John, 1928. "Zur Theorie der Gesellschaftsspiele," *Mathematische Annalen* 100, 295–320.

von Neumann, John, and Oskar Morgenstern, 1944. *Theory of Games and Economic Behavior*, Princeton University Press, Princeton.

Parsons, Charles, 1977. "What Is the Iterative Conception of Set?" in Robert Butts and Jaakko Hintikka (eds.), *Logic, Foundations of Mathematics and Computability Theory (Proceedings of the 5th International Congress of Logic, Methodology and Philosophy of Science 1975* 1), Dordrecht Reidel, Dordrecht, 335–367. Reprinted in Benacerraf and Putnam (1983), 503–529.

Partee, Barbara, 1984. "Compositionality," in F. Landman and F. Veltman (eds.), *Varieties of Formal Semantics*, Foris, Dordrecht, 281–312.

Pelletier, Francis Jeffry, 1994. "The Principle of Semantic Compositionality," *Topoi* 13, 11–24.

Poincaré, Henri, 1905–06. "Les mathematiques et la logique," *Revue de métaphysique et de morale* 13, 815–835 and 14, 17–34. English translation in *Science and Method*, Dover, New York, 1952, 143–159.

Prawitz, Dag, 1965, *Natural Deduction. A Proof-Theoretical Study*, Acta Universitatis Stockholmiensis, Stockholm.

Putnam, Hilary, 1971. *Philosophy of Logic*, Harper & Row, New York.

Rabin, Michael, 1957. "Effective Computability of Winning Strategies," in M. Dresher, A. W. Tucker, and P. Wolff (eds.), *Contributions to the Theory of Games* III (*Annals of Mathematics Studies* 39), Princeton Univrsity Press, 147–157.

Ramsey, Frank. P., 1925. "The Foundations of Mathematics," *Proceedings of the London Mathematical Society*, Ser. II, vol. 25, 338–384.

Ramsey, Frank P., 1930. "On a Problem of Formal Logic," *Proceedings of the London Mathematical Society*, Ser. 2, vol. 30, Part 4, 338–384.

Rantala, Veikko, 1975. "Urn Models: A New Kind of Non-Standard Model for First-Order Logic," *Journal of Philosophical Logic* 4, 455–474.

Rasiowa, Helena, and Roman Sikorski, 1963. *The Mathematics of Metamathematics*, Polska Akademia Nauk, Warsaw.

Russell, Bertrand, 1903. *The Principles of Mathematics*, Allen & Unwin, London.

Russell, Bertrand, 1905. "On Denoting," *Mind* 14, 479–493.

Russell, Bertrand, and A. N. Whitehead, 1910–13. *Principia Mathematica* I–III, Cambridge University Press, Cambridge.

Saarinen, Esa (ed.), 1979. *Game-Theoretical Semantics*, D. Reidel, Dordrecht.

Sandu, Gabriel, 1993. "On the Logic of Informational Independence and Its Applications," *Journal of Philosophical Logic* 22, 29–60.

Sandu, Gabriel, 1996. "IF First-Order Logic and Truth-Definitions," *Journal of Philosophical Logic*.

Sandu, Gabriel, and Jouko Väänänen, 1992. "Partially Ordered Connectives," *Zeitschrift für Mathematische Logik und Grundlagen der Mathematik* 38, 361–372.

Schnorr, Claus Peter, 1971. *Zufälligkeit und Wahrscheinlichkeit, (Lecture Notes in Mathematics* 218) Springer, Berlin.

Shapiro, Stewart, 1985. "Second-Order Languages and Mathematical Practice," *The Journal of Symbolic Logic* 50, 714–742.

Shapiro, Stewart, 1991. *Foundations without Foundationalism*, Clarendon Press, Oxford.

Simmons, Keith, 1990. "The Diagonal Argument and the Liar," *Journal of Philosophical Logic* 19, 277–303.

Simons, Peter, 1992. *Philosophy and Logic in Central Europe from Bolzano to Tarski*, Kluwer Academic, Dordrecht.

Skolem, Thoralf, 1922. "Einige Bemerkungen zur axiomatischen Begründung der Mengenlehre," *Proceedings of the Fifth Scandinavian Mathematics Congress*, Helsinki, 1922, 217–232.

Smullyan, Raymond, 1968. *First-Order Logic*, Springer, Berlin.

Stewart, Ian, 1992. *The Problems of Mathematics*, new edition, Oxford University Press, New York.

van Stigt, Walter P., 1990. *Brouwer's Intuitionism*, North-Holland, Amsterdam.

Szabo, M. E. (ed.), 1969. *The Collected Papers of Gerhard Gentzen*, North-Holland, Amsterdam.

Tarski, Alfred, 1933. "Projcie prawdy w jezykach nauk deducyjnych," *Prace Towarzystaw Naukowego Warszawsiego*, wydizial III, no. 34; English translation in Tarski 1956a; German translation as Tarski 1935.

Tarski, Alfred, 1935. "Der Wahrheitsbegriff in den formalisierten Sprachen," *Studia philosophica* 1, 261–405.

Tarski, Alfred, 1951. *A Decision Method for Elementary Algebra and Geometry*, 2nd ed. University of California Press, Berkeley and Los Angeles.

Tarski, Alfred, 1956a. "The Concept of Truth in Formalized Languages," in Tarski 1956b, 152–278.

Tarski, Alfred, 1956b. *Logic, Semantics, Metamathematics: Papers from 1923 to 1938*, Clarendon Press, Oxford.

Tarski, Alfred, 1959. "What Is Elementary Geometry?" in L. Henkin, P. Suppes, and A. Tarski (eds.), *The Axiomatic Method*, North-Holland, Amsterdam, 16–29.

Tarski, Alfred, 1992. "Drei Briefe an Otto Neurath," *Grazer Philosophische Studien* 43, 1–32.

Tennenbaum, Stanley, 1959. "Non-Archimedean Models for Arithmetic," *Notices of the American Mathematical Society* 6, 270.

Toepell, M.-M., *Über die Entstehung von David Hilberts "Grundlagen der Geometrie,"* Vandenhoek & Ruprecht, Göttingen, 1986.

Torretti, Roberto, 1978. *Philosophy of Geometry from Riemann to Poincaré*, D. Reidel, Dordrecht.

Troelstra, A. S., 1977. *Choice Sequences: A Chapter of Intuitionistic Mathematics*, Clarendon Press, Oxford.

Troelstra, A. S., and Dirk van Dalen, 1988. *Constructivism in Mathematics: An Introduction*, North-Holland, Amsterdam.

Tuuri, Heikki, 1990. *Infinitary Languages and Ehrenfeucht-Fraïsse Games*, Dissertation, University of Helsinki.

van der Waerden, B. L., 1985. *A History of Algebra*, Springer, Berlin.

Walkoe, W. Jr., 1970. "Finite Partially Ordered Quantification," *Journal of Symbolic Logic* 35, 535–555.

Wang, Hao, 1990. *Computation, Logic, Philosophy: A Collection of Essays*, Kluwer Academic, Dordrecht.

Westerstahl, Dag, 1989. "Quantifiers in Formal and Natural Languages," in D. Gabbay and F. Guenther (eds.), *Handbook of Philosophical Logic* IV, Reidel, D., Dordrecht, 1–131.

Weyl, Hermann, 1917. *Das Kontinuum*, Veit, Leipzig.

Williams, Neil H., 1977. *Combinatorial Set Theory*, North-Holland, Amsterdam.

Wittgenstein, Ludwig, 1922. *Tractatus Logico-Philosophicus*. The German text of Ludwig Wittgenstein's *Logisch-philosophische Abhandlung* with English translation by C. K. Ogden, Routledge & Kegan Paul, London.

Youschkevitch, A. P., 1976. "The Concept of Function up to the middle of the 19th century," *Archive for the History of Exact Sciences* 16, 37–85.

Zermelo, Ernst, 1908. "Untersuchungen über die Grundlagen der Mengenlehre I," *Mathematische Annalen* 65, 261–281.

Index of Names

Ackermann, Wilhelm, 5, 6, 46, 201, 274
Anderson, C. Anthony, 273
Archimedes, 8
Aristotle, 3, 21, 123, 164
Aspray, William, 276

Baire, René, 185
Baldus, Richard, 92, 271
Bar-Hillel, Yehoshua, 42
Barwise, Jon, 12, 56, 71, 121, 185, 190, 271
Beltrami, Eugenio, 164
Benacerraf, Paul, 21, 210, 271, 273, 274, 277
van Benthem, Johan, 271, 275
Beth, Evert W. 6, 61
Blass, Andreas, 198, 271
Blumenthal, Otto, 21, 271
Boltzmann, Ludwig, 21, 271
Bolzano, Bernhard, 189
Boole, George, 2, 166, 258, 263
Boolos, George, 168, 271
Borel, Emile, 226
Bottazzini, Umberto, 253, 271
Brouwer, Luitzen E. J., 237, 238, 252
Butts, Robert, 277

Cantor, Georg, 2, 164–169, 181, 182, 273
Carnap, Rudolf, x, xi, 12, 16, 17, 152
Cauchy, Augustin-Louis, 9, 273
Chomsky, Noam, 6, 46, 54, 126

Church, Alonzo, 12, 18, 114
Cohen, Paul, 272
Cummins, Robert, 120, 272

van Dalen, Dirk, 237, 279
Dancy, Jonathan, 274
Dauben, Joseph W., 164, 168, 182, 272
Davidson, Donald, 66, 106, 272
Dawar, Anuj, 93, 272
Devlin, Keith, 272
Dedekind, Richard, 200
DeMorgan, Augustus, 73, 133, 147
Desargues, Girard, 21
Descartes, René, 24, 89
Dreben, Burton, xi
Dresher, M., 277
Dummett, Michael, 22, 26, 27, 36, 38, 39, 211, 212, 215, 218, 222, 244, 245, 272

Easton, William B., 272
Ebbinghaus, Heinz-Dieter, 15, 141, 272
Ekeland, Ivar, 229, 272
Enderton, H. B., 56, 71, 121, 272
Erdös, Paul, 77, 78, 200, 272
Euclid, 1, 2, 91, 164
Euler, Leonhard, 253, 271, 272

Feferman, Solomon, 12, 185, 201, 202, 266–268, 271, 272

Index prepared by Risto Vilkko

Felgner Ulrich, 272
Fenstad, Jens-Erik, 33, 147, 181
Fernández Moreno, Luis, 130, 272
Fine, Terrence, 228, 272
Flum, Jörg, 15, 141
Fraenkel, Adolf, 163, 176, 202
Freeman, Eugene, 274
Frege, Gottlob, vii, 2, 5, 11, 12, 18, 46–
    51, 54–56, 66, 67, 69, 71, 90, 94,
    106, 125, 126, 161, 164, 181, 182,
    186, 195, 200, 207,
Freud, Sigmund, 252
Freudenthal, Hans, 92, 273

Gabbay, Dov M., 279
Gaifman, Haim, 273
Gavroglu, Kostis, 275
Georg, Alex, 274
Gödel, Kurt, xi, 6, 14–18, 30, 31, 41,
    88–91, 93–99, 102, 113–119, 131,
    132, 141–145, 150–152, 165, 166,
    168, 169, 172–177, 199, 204, 222,
    223, 225, 229, 231–235, 259–261,
    265, 266, 273
Goldbach, C., 192, 197
Goldfarb, Warren, 210, 273
Grabiner, Judith V., 74, 273
Graham, Ronald, L., 78, 200, 273
Grattan-Guinness, Ivor, 277
Guenther, Franz, 279
Gurevich, Yuri, 198, 271

Hadamard, Jacques, 244
Hales, A. W., 78, 79
Hallett, Michael, 168, 273
Hegel, Georg W. Friedrich, 211
Heim, Irene, 157, 273
Hella, Lauri, 93
Henkin, Leon, 51, 56, 79, 98, 102, 117–
    119, 121, 165, 192, 193, 249, 263,
    271, 273, 278
Heyting, Arend, viii, 212, 246–249, 271,
    273
Hiipakka, Janne, xi
Hilbert, David, xi, 1–6, 8–11, 16, 19, 21,
    40, 46, 69, 89–92, 98, 164, 182,
    199–201, 206, 208, 219, 222, 233,
    234, 271, 274
Hilpinen, Risto, 28, 212, 274
Hintikka, Jaakko, 12, 13, 17, 20–23, 27,
    29, 36, 44, 52, 66, 68, 73, 79, 82,

    83, 103, 104, 109, 110, 124–127,
    139, 157, 160, 165, 193, 199, 208,
    223, 224, 239, 251, 274, 275, 277
Hintikka, Merrill B., 17, 23, 127, 275
Hölder, (Ludwig) Otto, 204
Horn, Laurence R., 275
Husserl, Edmund, 108, 120, 196
Hughes, R. I. G., 77, 275
Huuskonen, Taneli, 270

Jackendorff, Ray S, 275
Jefferson, Thomas, vii
Jeroslow, R. G., 275
Jewett, R. I., 78, 79
Jones, James, P., 28, 219, 275

Kanamori, Akihiro, 169, 276
Kant, Immanuel, 3, 89, 104
Keenan, Edward, 57, 79, 276
Kino, A., 276
Kitcher, Philip, 276
Kleene, Stephen, C., 235, 258, 259, 276
Klein, Felix, 164
Klein, Esther, 78
Kleinberg, Eugene M., 200, 276
Kolaitis, Phokion G., 130
Körner, Stephan, 274
Kreisel, George, 40, 217, 276
Kripke, Saul, xi, 122, 161, 254, 259,
    266–268, 270, 276
Krynicki, Michael, 117, 118, 188, 276
Kuhn, Thomas, vii
Kunen, Kenneth, 276
Kulas, Jack, 27, 44, 109, 110, 125, 157,
    275
Kusch, Martin, 276

Landman, Fred, 277
Leibniz, Gottfried W., 112, 122, 125
Lenin, Vladimir I., vii
LePore, Ernest, 274
Lesniewski, Stanislaw, 17, 108
Lindström Per, 18, 141, 276
Lovejoy, Arthur O., 168, 276
Löwenheim, Leopold, 6, 18, 59, 141, 189,
    191, 202, 203
Luosto, Kerkko, 270

Mach, Ernst, 21
Magidor, Menachem, 169, 276
Mancosu, Paolo, 246, 276

Martin, Robert L., 131, 161, 276
Martin-Löf, Per, 228, 276
Matiyasevich, Yuri V., 219, 276
Maxwell, James Clerk, 2
McGee, Vann, 173, 276
Mendelson, Elliott, 114, 143, 276
ter Meulen, Alice, 271, 275, 276
Moore, Gregory, 170, 195, 201, 244, 276, 277
Morgenstern, Oskar, 277
Morris, Charles, 42, 277
Moschovakis, Yannis N., 210, 277
Mostowski, Andrzej, 15, 40, 217, 277
Müller, Gert H., 276
Myhill, John, 276

von Neumann, John, 24, 79, 226, 277
Neurath, Otto, 17, 204

Ogden, C. K., 279
Oikkonen, Juha, 122, 275

Parsons, Charels, 168, 277
Partee, Barbara, 106, 277
Pascal, Blaise, 21
Peano, Giuseppe, vii, 98, 153, 197, 202, 223, 259, 265, 267, 268
Peirce, Charles S., 28, 46, 212, 274
Pelletier, Francis J., 106
Poincaré, Henri, 182, 277
Prawitz, Dag, 212, 277
Puhl, Klaus, 275
Putnam, Hilary, 16, 21, 271, 274, 277

Quine, W.V., 7, 12, 18, 67, 129, 228

Rabin, Michael, 219, 277
Ramsey, Frank, 77–79, 137, 165, 194, 200, 210, 273, 277
Rantala, Veikko, 12, 29, 103, 160, 239, 275, 277
Rasiowa, Helena, 203, 277
Remes, Unto, 104, 275
Rorty, Richard, 130
Rotschild, Bruce L., 78, 200, 273
Russell, Bertrand, vii, ix, x, xii, 5, 12, 43, 46, 47, 49–51, 55, 66, 67, 94, 161, 169, 174, 180, 186, 194, 195, 210, 273, 277

Saarinen, Esa, 274, 278
Sandu, Gabriel, xi, 27, 52, 56, 79, 82, 110, 122, 126, 224, 254, 275, 278

Savage, C. Wade, 273
Schnorr, Claus Peter, 228, 278
Schrödinger, Erwin, 2
Schwabl, H.-D, 275
Scott, Dana, 276
Shapiro, Stewart, 192, 193, 278
Shoenfield, Joseph, 168
Sikorski, Roman, 203, 277
Simmons, Keith, 143, 278
Simons, Peter, 278
Skolem, Thoralf, 6, 18, 30, 34, 39, 48, 49, 51, 59, 60, 79, 80, 141, 174, 175, 177, 179, 189, 191, 198, 202, 203, 213, 217, 221, 223, 224, 228, 236, 242, 246, 248, 249, 251, 278
Smullyan, Raymond, 68, 278
Sosa, Ernst, 274
Spencer, Joel H., 78, 200, 273
Stewart, Ian, 29, 30, 44, 278
van Stigt, Walter P., 278
Szabo, M. E., 278
Suppes, Patrick, 278
Szekeres, George, 77, 78

Tarski, Alfred, xi, 13–19, 22, 28, 29, 32, 41, 96–99, 101, 105–108, 110–112, 116, 117, 120–124, 128–131, 133, 138–141, 144, 171–174, 182, 204, 266, 278
Tennenbaum, Stanley, 223, 278
Teubner, B. G., 274
Thomas, Wolfgang, 15, 141
Toepell, Michael-Markus, 21, 278
Torretti, Roberto, 164, 279
Troelstra, Anne S., 237, 245, 279
Turing, Alan, 95, 215, 237
Tucker, A. W., 277
Tuuri, Heikki, 130, 279

Väänänen, Jouko, 56, 122, 188, 270, 275, 276, 278
Veltman, Frank, 277
Versley, Richard E., 276

van der Waerden, B. L., 204, 279
Walkoe, Wilbur John Jr., 56, 71, 121, 279
Wang, Hao, 168, 198, 279
Weierstrass, Karl, 29, 44, 189, 271
Westerstahl, Dag, 57, 79, 276, 279
Weyl, Hermann, 179, 279

Williams, Neil H., 200, 279
Wittgenstein, Ludwig, 12, 18, 22, 23, 32, 42, 44, 127, 162, 211, 215, 222, 226, 279
Whitehead, Alfred N., vii, xii, 46, 180, 194, 277

Wolff, P., 277

Youschkevitch, A. P., 251, 279

Zermelo, Ernst, 251, 279

Index of Subjects and Titles

analysis, conceptual, 9
arithmetic, 79
  elementary, viii, 14, 30, 89, 90, 93–98, 101, 113, 116, 118, 143, 145, 171, 197, 206, 223, 224, 233, 235
  first-order (IF), 132
  Peano, 98, 197, 202, 223, 259, 265, 267, 269
  undefinability of 15
axiom
  Archimedean, 8
  Aussonderung (Zermelo's), 179
  of choice, ix, 7, 19, 32, 40–42, 80, 82, 83, 165, 175, 176, 193, 201, 213, 220–223, 244
  deductive, viii
  of determinacy, 33
  of geometry, 8
  of induction, 153
  Kripke-Feferman, 268
  of large cardinals, 169
  of choice, ix, 7, 19, 32, 40–42, 80, 82, 83, 165, 175, 176, 193, 201, 213, 220–223, 244
  of completeness, 8, 92, 208
  of comprehension, 165, 176, 178, 195
  of constructibility, 168
  of continuity, 3
  Peano, 153
  of reducibility (Russell's), 180
  of set existence, 194
axiomatization, 1, 9, 65–68, 88–104, 126, 145, 146, 183, 251, 266
  of arithmetic, 94
  complete, viii, 6, 65–67, 153, 184, 224, 249
  of geometry (Euclidean), 2
  of geometry (Hilbert's), 2, 3, 8, 10, 11
  of logic, 4, 5
  nonlogical, 1, 3, 4,

Begriffsschrift, 11, 125

Church's Thesis, 114
completeness, x, 1, 19, 67, 88–104, 145, 163, 166, 169, 170, 192, 233
  deductive, 91, 92
  descriptive, 91
  Hilbertian, 92
  semantical, 91
compositionality, 13, 106–112, 116, 118, 121, 139, 140, 149
  principle of, ix
conjectures, 193
  Goldbach's, 192, 197
constructivism, ix, x, 7, 38, 39, 121, 137, 142, 161, 177, 210–237, 241, 243–245, 250

Index prepared by Risto Vilkko

continuity, 9, 10
continuum, 193
course-of-values (Frege's *Werthverläufe*), 181
classes, viii, 12

Definition $\varepsilon - \delta$, 9, 29
dependence vs. independence, x, 72–87,
   of quantifiers, 47–53, 56–58, 70
differentiation, 9
distinction
   de dicto vs. *de re*, 72, 124
   definitory vs. strategic rules, 128
   logical vs. mathematical reasoning, 90,
   semantics vs. pragmatics, 42

equations, 2, 185
equicardinality, viii, 186, 187, 190, 191

finiteness, 7
functions, 40, 193, 253
   in extension (Russell), 174
   Skolem, 30, 31, 34, 35, 39, 48, 49, 51,
     60, 79, 80, 174, 175, 177, 213, 217,
     221, 223, 224, 228, 236, 242, 246,
     248, 249, 251

geometry, 9, 10, 78, 89, 97, 234
   applied, 3
   elementary, 197
   Euclidean, 1, 2, 164
   nonEuclidean, 97, 164
   Grundlagen der Geometrie, 1, 4, 10, 92

Impossibility result (Tarski's), 15, 22,
   123, 171, 172
induction
   mathematical, viii, 7, 184, 188, 191
   transfinite, 189
inference, 4, 5, 20, 21, 33, 37, 71, 88
   Hilbert's, 8
   modes of, viii, 184, 190
infinity, viii, 184, 187
interpretation, 3, 4
   constructivistic, ix, 224, 225, 235, 251
   epistemic, 242–246, 250–253
   Gödel's functional (Dialectica), 222, 223, 232, 233, 235
   intuitionistic, 247

standard vs. nonstandard, 165, 166,
   192–196, 201, 210, 213, 246
intuitionism, 134, 161, 177, 229,
   237–247, 252

knowledge of objects vs. knowledge of
   facts, x, 86, 252

language games, 32–39, 159, 212, 229,
   238
   Wittgensteinian, 22, 23, 42, 211, 222
law
   de Morgan's, 73, 133, 147
   of double negation, 73, 133, 147
   of the excluded middle, *see tertium non datur*
lemma, diagonal, 131, 142–145, 150,
   152, 172–174, 180
logic
   epistemic, 83–87, 238–253
   functions of, 1–21
   deductive/descriptive, 4, 9, 10, 19, 20,
     46, 92, 149, 209, 210, 230–234
   higher-order, 6, 7, 16, 97, 104, 126,
     163, 165, 186, 190, 192, 194, 195,
     201, 203, 207–209, 223, 231
   independence-friendly (IF), 46–163,
     178–210, 217, 229, 241, 247, 254–
     270
   intuitionistic, viii, 134, 212, 238, 246–
     249
   modal, 5
   model-theoretical, 12, 13
   ordinary first-order, viii, ix, 6, 7, 9,
     11, 16, 18, 33, 34, 36, 43, 46–51,
     54–61, 65–70, 73, 74, 77, 82, 89,
     90, 93, 94, 96, 109, 125–129, 132,
     135, 138, 141, 143, 152, 163, 170,
     171, 174, 184–190, 194, 195, 198,
     201–205, 209, 213, 216, 217,223,
     229
   second-order, 7, 65, 89, 98, 129, 150,
     191–197, 213, 220, 225
   syllogistic, 89
logicism, 183, 190

method
   analytic, 89
   axiomatic, 2, 10, 94, 206, 234
   deductive, 100
   mathematical, 89
   semantical vs. syntactical, 90

model
  concept of, 12
  interrogative, 36, 39
  urn, 12, 103
model theory, viii, 11–13, 16, 19–21, 85,
    90, 102, 104, 129, 141, 152, 162,
    163, 165, 170–174, 184, 192, 197,
    239, 249

nominalism, 7, 117, 198, 220, 234, 250
numbers
  natural, 8, 102, 104, 116, 166, 177, 184,
    206, 207, 217, 219
  real, 10, 102, 177, 184, 185, 189
  Gödel, 14, 30, 113–119, 131, 132, 142,
    143, 145, 151, 152, 172–174, 180,
    225, 259–261, 265, 266

paradox, 7, 97, 101, 121, 145, 165, 167,
    175, 239
  liar, 131, 132, 142–144, 150, 152, 159,
    160, 176
  of logical omniscience, 239
  Meno's, 34, 35, 38, 39
  Skolem, 202, 203
possible worlds, 12, 34, 66, 239, 240
predicates, viii, 6, 185–188
Principia Mathematica, vii, x
Principles of Mathematics, vii, ix, x
proof, 5, 10, 38

relation, viii, 4, 6, 8, 113–115, 135, 188,
    194
  epsilon, 115, 135, 188, 194, 178

semantical games, 25–27, 31–45, 57, 58,
    63, 70, 115, 119–123, 128, 132,
    134, 137, 138, 147, 148, 154, 157,
    158, 174, 211–219, 227, 228, 231,
    235–238
semantics
  game-theoretical (GTS), 25–50, 54,
    57, 58, 72, 77, 90, 103, 109, 112,
    115, 119–122, 125, 132–134, 148,
    157, 160, 163, 201, 210, 211,
    213–222, 225–227, 230, 232, 236,
    241, 243, 254–270
  ineffability of, 17, 18
  logical, 11
  truth-conditional, 23
  verificationist, 23

set
  power, viii, 7, 184, 188, 191
  Hintika, 68
set theory, viii, 3, 8, 16, 33, 89, 97, 99,
    129, 132, 152, 185, 186, 189, 191,
    198–207, 213
  axiomatic, ix, 7, 18, 19, 74, 100, 102,
    104, 126, 163–182, 185, 191–195,
    202, 207, 222
slash-notation (/), 51–53, 254
Sprachlogik, 46, 55, 124, 126

tertium non datur, ix, 32, 33, 65, 68,
    73, 131–146, 161, 178, 179,
    181, 212, 217, 219, 258,
    267
theorem
  Beth's, 7, 61
  Bolzano–Weierstrass, 189
  deduction, 149
  Desargues', 21
  Godel's incompleteness, 18, 93, 95,
    96, 132, 165, 166, 173
  Gödel's impossibility, 142
  Hales–Jewett, 78, 79
  interpolation, 6
  Lindström, 18, 141
  Löwenheim–Skolem, 6, 18, 59, 141,
    191, 202, 203
  Pascal's, 21
  Ramsey-type, 137
  separation, 6, 61, 133
  Tarski's, 16
theory
  elementary number, 153
  game, see semantics, GTS
  group, 9, 10, 167
  lattice, 3, 9, 10, 167
  model, see model theory
    of anaphora (GTS), 157, 158
    of fields, 9
    of finite types, 197
    of reals, 197
    of types (Russell's), 194
  proof, 11, 141, 162, 234
  quantum, 76, 77
  Ramsey, 77–79, 200
  set, see set theory
  type, 205, 207
Tractatus Logico-Philosophicus,
    153

truth definition, viii, ix, 1–45, 105–130, 144
Tarski's, 13–18, 22, 28, 29, 32, 65, 105–112, 120–124, 128, 129, 133, 266
truth definition, (cont.)
  recursive, 14
  game-theoretical, 26, 27, 30, 32, 34, 35, 41, 119–122, 172–174, 220, 221
  pragmatist, 44

T-schema (Tarski's), 41, 106, 138–140, 172–174

universal language, 17–19, 204, 205

Vienna Circle, 17, 18, 233

Warsaw School, 17
well-ordering, viii, 7, 184, 188, 193